Raman Scattering on Emerging Semiconductors and Oxides

Raman Scattering on Emerging Semiconductors and Oxides presents Raman scattering studies. It describes the key fundamental elements in applying Raman spectroscopies to various semiconductors and oxides without complicated and deep Raman theories.

Across nine chapters, it covers:

- SiC and IV-IV semiconductors,

- III-GaN and nitride semiconductors,

- III-V and II-VI semiconductors,

- ZnO-based and GaO-based semiconducting oxides,

- Graphene, ferroelectric oxides, and other emerging materials,

- Wide-bandgap semiconductors of SiC, GaN, and ZnO, and

- Ultra-wide gap semiconductors of AlN, Ga_2O_3, and graphene.

Key achievements from the author and collaborators in the above fields are referred to and cited with typical Raman spectral graphs and analyses.

Written for engineers, scientists, and academics, this comprehensive book will be fundamental for newcomers in Raman spectroscopy.

Zhe Chuan Feng has had an impressive career spanning many years of important work in engineering and tech, including as a professor at the Graduate Institute of Photonics & Optoelectronics and Department of Electrical Engineering, National Taiwan University, Taipei; establishing the Science Exploring Lab; joining Kennesaw State University as an adjunct professor, part-time; and at the Department of Electrical and Computer Engineering, Southern Polytechnic College of Engineering and Engineering Technology. Currently, he is focusing on materials research for LED, III-nitrides, SiC, ZnO, other semiconductors/oxides, and nanostructures and has devoted time to materials research and growth of III-V and II-VI compounds, LED, III nitrides, SiC, ZnO, GaO, and other semiconductors/oxides.

Professor Feng has also edited and published multiple review books in his field, alongside authoring scientific journal papers and conference/proceeding papers. He has organized symposiums and been an invited speaker at different international conferences and universities. He has also served as a guest editor for special journal issues.

Raman Scattering on Emerging Semiconductors and Oxides

Dr Zhe Chuan Feng

CRC Press
Taylor & Francis Group
Boca Raton London New York

CRC Press is an imprint of the
Taylor & Francis Group, an **informa** business

Designed cover image: Shutterstock_1895444176

First edition published 2025
by CRC Press
2385 NW Executive Center Drive, Suite 320, Boca Raton FL 33431

and by CRC Press
4 Park Square, Milton Park, Abingdon, Oxon, OX14 4RN

CRC Press is an imprint of Taylor & Francis Group, LLC

© 2025 Zhe Chuan Feng

ISBN: 9781032638874 (hbk)
ISBN: 9781032644899 (pbk)
ISBN: 9781032644912 (ebk)

DOI: 10.1201/9781032644912

Typeset in Minion Pro
by Apex CoVantage, LLC

Contents

Preface

SINCE THE DISCOVERY OF Raman scattering in 1928, nearly one century ago, especially after the invention of the laser in the 1960s, Raman applications in semiconductors, oxides, and other fields have rapidly and widely developed, accompanied by various powerful Raman instruments greatly developed and improved. Raman laser spectroscopy can be used to diagnose, characterize, and analyze matter and is widely applied in various sciences and engineering, such as physics, chemistry, materials, biology, medicine, environment science, and engineering. Research and development (R&D) in semiconductors and industry have been greatly developed attractively in the past several decades and are continually enhancing throughout the world range. Raman spectroscopy has become an important and useful characterization technology in the investigation of solids, semiconductors, and oxides and has promoted the R&D of science, technology, industry, and economy.

The book, *Raman Scattering on Emerging Semiconductors and Oxides*, is going to meet the demands of these fields. The author, Zhe Chuan Feng, has devoted himself to the R&D of semiconductors for over 40 years and employed the Raman spectroscopic technique in the characterization and investigation of various semiconductors and oxides, since the 1980s, with fruitful research accommodation and outputs. This book aims to provide the fundamentals of Raman scattering and typical and practical Raman scattering studies, with nine chapters on SiC and IV-IV semiconductors, III-nitride semiconductors, III-V and II-VI semiconductors, ZnO- and GaO-based semiconducting oxides, graphene, ferroelectric, and other emerging materials. The book describes the key elements in applying Raman spectroscopies to various semiconductors and oxides without complicated and deep Raman theories but with necessary and brief items. All materials covered are based on the research outputs from the author and collaborators in the past four decades. These results provide clear evidence that Raman laser spectroscopy is a convenient and powerful tool for the diagnostics of various semiconductors and oxides. Because of the limit of pages, more studies on some materials and structures are referred to with references only. Many practical Raman spectral graphic examples are presented and demonstrated. Multiple and special Raman techniques and skills, including wavelength-dependent, temperature-dependent, angle-dependent, pressure-dependent, and resonant Raman scattering, are described and discussed. Raman spectral line shape analyses, Gaussian and Lorentzian fit, the space correlation model, and stress and strain calculation are introduced, which are useful to analyze and evaluate the material crystalline quality of semiconductors and oxides for industry production and investigation.

The book is written to serve industry and academic professionals, engineers, scientists, professors, teachers, graduates, upper undergraduates, and newcomers. The current book with nine chapters elaborating on many semiconductors and oxides can be an excellent practical guide in the field and will be useful to engineers, scientists, students, and newcomers. It is greatly helpful for them to use Raman spectroscopy and to promote the Raman scattering studies, applications, and developments of many important and advanced semiconductors, oxides, and related fields.

Acknowledgments

I'D LIKE TO ACKNOWLEDGE Profs./Drs. Wolfgang J. Choyke, Sidney Perkowitz, Ian. T. Ferguson, Devki N. Talwar, Andrew T. S. Wee, Sing Hai Tang, Soo Jin Chua, Gu Xu, Zexiang Shen, Ajeet Rohatgi, A. Erbil, Kuo To Yue, R. A. Stall, Jim B. Webb, Chin-Che Tin, Petre Becla, K. P. Williams, G. D. Pitt, Jian H. Zhao, A. Mascarenhas, Tzuen-Rong Yang, Y. T. Hou, Frederick H. Long, J. C. Burton, Xiong Zhang, Chih-Chung Yang, Hao-Hsiung Lin, Li-Chyong Chen, Kuei-Hsien Chen, W. Tong, R. P. G. Karunasiri, Lingyu Wan, Wenhong Sun, Lianshan Wang, Kaiyan He, Yao Liu, Changtai Xia, Weijie Lu, Shu-De Yao, Xiaodong Hu, Dong-Sing Wuu, Ray-Hua Horng, Na Lu, Wei Zheng, Zhi Ren Qiu, Ting Mei, Deng Xie, Shuai Chen, Qingxuan Li, Hanling Long, Feng Wu, Jiangnan Dai, Changqing Chen, Jianwei Ben, Xiaojuan Sun, Dabing Li, Jianguo Zhao, Shuchang Wang, Xuguang Luo, Bin Xin, Weijie Lu, Na Lu, Nikolaus Dietz, Jeffrey Yiin, Benjamin Klein, among others. Special thanks to my many collaborators and past students. With all their help and support, this book was successfully completed.

Finally, besides the acknowledgments to many professors, collaborators, and my past students, listed at the beginning of the book, I would like to thank the editors and colleagues from the CRC Press/Taylor & Francis Group and their Production Division, for their great help and support that guaranteed the success of this book.

About the Author

D r Zhe Chuan Feng earned his PhD in condensed matter physics from the University of Pittsburgh, Pennsylvania, USA in 1987 and, earlier, BS (1962–68) and MS (1978–81) from the Department of Physics at Peking University, Beijing, China. He worked at Emory University (1988–92); National University of Singapore (1992–94), Georgia Tech (1995); EMCORE Corporation (1995–97); Institute of Materials Research & Engineering, Singapore (1998–2001); Axcel Photonics (2001–02); Georgia Tech (2002–03); National Taiwan University (NTU) (2003–2015.1) as a professor at Graduate Institute of Photonics & Optoelectronics and Department of Electrical Engineering; and Guangxi University (GXU) (2015–2020) as a distinguished professor at School of Physical Science and Technology. After he got double-retired from NTU and GXU, back home in Georgia, USA, he established the Science Exploring Lab, and in January 2022, he joined Kennesaw State University, as an adjunct professor, part-time, in the Department of Electrical and Computer Engineering, Southern Polytechnic College of Engineering and Engineering Technology, currently focusing on materials research for LED, III-nitrides, SiC, ZnO, GaO, Graphene, other semiconductors/oxides, and nanostructures.

Zhe Chuan Feng is a well-known expert in the field. He has devoted much time to materials research and growth of III-V and II-VI compounds, LED, III-nitrides, SiC, ZnO, GaO, and other semiconductors/oxides. Professor Feng has edited/published 12 review books on compound semiconductors and microstructures; porous Si, SiC, and III-nitrides; ZnO devices; and nanoengineering, especially in the 21st century on wide-bandgap semiconductors: *SiC Power Materials – Devices and Applications*, Springer (2004); *III-Nitride Semiconductor Materials* (2006) and *III-Nitride Devices and Nanoengineering* (2008), Imperia College Press; *Handbook of Zinc Oxides and Related Materials: Volume 1 – Materials, and Volume 2 – Devices and Nano-Engineering* (2012) and *Handbook of Solid-State Lighting and LEDs* (2017), T&F/CRC; *III-Nitride Materials, Devices and Nanostructures*, World Scientific Publishing (2017); and *Handbook of Silicon Carbide Materials and Devices*, T&F/CRC (2023).

From Google Scholar, https://scholar.google.com/citations?hl=en&user=vdyXZpEAA AAJ, he has citations >7800, h-index:**42**, i10-index:**170**, from > 520 SCI cited papers with many on Raman scattering for a variety of semiconductors, oxides, and related materials. He has authored and co-authored >430 scientific journal papers and >420 conference/proceeding papers, among which there are >90 journal papers and >50 conference papers on Raman scattering or using Raman spectroscopy. He has been a symposium organizer and invited speaker at different international conferences and universities. He has served as a guest editor for special journal issues and has been a guest professor at Sichuan University, Nanjing Tech University, South China Normal University, Huazhong University of Science & Technology, Nankai University, and Tianjin Normal University. Professor Feng has been a fellow of SPIE since 2013. More details about his academic contributions can be seen on the previous web link and at the web of NTU for retired professors: www.ee.ntu.edu.tw/profile1.php?teacher_id=941011&p=5.

Zhe Chuan Feng, has engaged in the research and development (R&D) of semiconductors for more than 40 years and employed Raman spectroscopic techniques in the characterization and investigation of various semiconductors and oxides, since the 1980s, with fruitful research accommodation and outputs. He has edited/published 12 review-style books on compound semiconductors and microstructures; porous Si, SiC, and III-nitrides; ZnO devices; and nanoengineering. He has authored and co-authored >90 journal papers and >50 conference papers on Raman scattering, occupying a big ratio in his scientific outputs.

12 PUBLISHED BOOKS EDITED BY ZHE CHUAN FENG (2 CO-ED.)

<<**Handbook of Silicon Carbide and Related Materials**>>, Editor: Zhe Chuan FENG, 15-chapters, CRC Press, Taylor & Francis Group, London/New York, 30 May 2023. ISBN 9780367188269. www.routledge.com/9780367188269

<<**Handbook of Solid-State Lighting and LEDs**>>, Editor: Zhe Chuan FENG, 24-chapters, 705-pages, CRC press, Taylor & Francis Group, London/New York, 2017, ISBN 9781498741415. *https://doi.org/10.1201/9781315151595*.

<<**III-Nitride Materials, Devices and Nanostructures**>>, Editor: Zhe Chuan FENG, 410-pages, World Scientific Publishing Press, Singapore, 2017, ISBN: 978-1-78634-318-5. https://doi.org/10.1142/q0092.

<<**Handbook of Zinc Oxides and Related Materials:** *Volume 1) Materials, and Volume 2) Devices and Nano-Engineering*>>, Editor: Zhe Chuan FENG, 440pp+640pp, CRC press, Taylor & Francis Group, London/New York, 2013, ISBN 9781439855706. DOI: 10.1201/b13068; https://doi.org/10.1201/b13072.

<<**III-Nitride Devices and Nanoengineering**>>, Editor: Zhe Chuan FENG, 462pp, Imperia College Press, London, 2008, ISBN: 978-1-84816-223-5.

<<**III-Nitride Semiconductor Materials**>>, Editor: Zhe Chuan FENG, 440pp, Imperia College Press, London, 2006, ISBN: 978-1-86094-636-3.

<<**SiC Power Materials – Devices and Applications**>>, Editor: Zhe Chuan FENG, Springer, Berlin, 450 pp. 2004, ISBN: 978-3-642-05845-5.

<<**Silicon Carbide: Materials, Processings and Devices**>>, Editors: Zhe Chuan FENG and Jian H. ZHAO, Taylor & Francis Books, Inc., New York, 416pp, 2003, ISBN: 9781591690238.

<<**Porous Silicon**>>, Editors: Zhe Chuan FENG and Raphael TSU, 488pp, World Scientific Publishing, Singapore, 1994, ISBN: 978-981-02-1634-4.

<<**Semiconductor Interfaces, Microstructures and Devices: Properties and Application**>>, Editor: Zhe Chuan FENG, 308pp, CRC Publisher, London, 1993, ISBN:0750301805.

<<**Semiconductor Interfaces and Microstructures**>>, Editor: Zhe Chuan FENG, World Scientific Publishing, Singapore, 328pp, 1992, ISBN: 978-981-02-0864-6.

Book Summary

IN SUMMARY, THIS BOOK presents Raman scattering studies in nine chapters covering SiC and IV-IV semiconductors, III-GaN and nitride semiconductors, III-V and II-VI semiconductors, ZnO-based and GaO-based semiconducting oxides, graphene, ferroelectric oxides, and other emerging materials. Wide-bandgap semiconductors of SiC, GaN, and ZnO and ultra-wide gap semiconductors of AlN, Ga_2O_3, and graphene are emphasized, in addition to traditional III-V and II-VI semiconductors. Key achievements from the author and collaborators in the aforementioned fields are referred to and cited with typical Raman spectral graphs and analyses.

This book possesses outstanding features: it focuses on Raman scattering for emerging and key semiconductors and oxides nowadays important in research and industry. It deals with widegap SiC, III-nitride, and ZnO-based semiconductors and emphasizes ultra-wide energy gap materials of AlN ($E_g = 6.2$ eV) and Ga_2O_3 (E_g near 5 eV). It exhibits typical Raman studies for traditional II-VI and III-V semiconductors, graphene, ferroelectric oxides, and others. It presents plenty of graphs and figures helpful in understanding Raman scattering on key semiconductors and oxides. It acts as a practical guide for Raman scattering studies and applications. It is intended for industry and academic professionals, engineers, students, and newcomers to help in using Raman spectroscopy.

More practical Raman spectral graphic examples are present and demonstrated. Multiple and special Raman techniques and skills, including wavelength-dependent, temperature-dependent, angle-dependent, pressure-dependent, and resonant Raman scattering, are described and discussed. Raman spectral line shape analyses, Gaussian and Lorentzian fits, the space correlation model, and stress and strain calculation are introduced, which are useful to analyze and evaluate the material crystalline quality of semiconductors and oxides for industry production and investigation. This book describes the key fundamental elements in applying Raman spectroscopies to various semiconductors and oxides without complicated and deep Raman theories. The book is intended to serve industry and academic professionals, engineers, scientists, professors, teachers, graduates, and upper undergraduate students, and so on, and would be greatly helpful to newcomers to use Raman spectroscopy.

I would be pleased if readers in various fields could take this book as their practical guide to learn and handle Raman spectroscopies and develop further in their own fields; some materials, such as boron nitride, are not included yet.

Zhe Chuan FENG
Georgia, USA

CHAPTER 1

Introduction

1.1 FUNDAMENTALS OF RAMAN SCATTERING

Raman scattering (RS) describes the interaction between light and material as a light bean incident on material. Sir Chandrasekhara Vankata Raman in 1928 found that the light scattered by liquids such as benzene contains sharp sidebands symmetrically disposed around the incident frequency with shifts identical to the frequencies of some infrared vibrational lines [1.1]. Indeed, this type of inelastic scattering of light was predicted by Adolf Smekal theoretically in an article on Dispersion Quantum Theory in 1923 [1.2], but the experimental observation was delayed till the discovery of C. V. Raman and K. S. Krishnan in 1928, nearly one century ago [1.1]. At a similar time, G. S. Landsberg and L. I. Mandelstam independently found this phenomenon in quartz [1.3]. As a certain wavelength light is incident onto a sample, no matter whether the sample is in a solid, liquid, gaseous, or other phase, the incident light beam will be disturbed by molecular or atomic vibrations from detected materials, leading to the appearance of new wavelength components in the outgoing light spectra.

This type of inelastic scattering phenomenon was named Raman scattering, and C. V. Raman won a Nobel Prize in Physics in 1930. It is caused by the modulation of the susceptibility (or, equivalently, polarizability) of the medium by the vibrations. Different materials possess their different and unique Raman spectral lines, which are also influenced by environmental conditions, such as temperature, pressure, and radiation. However, Raman scattering signals are usually very weak. It is necessary to have a monochromatic light source with strong power focused on a small beam size, sensitive weak signal detection electronics, and a data processing system. Due to these high technical requirements, investigation and applications of Raman scattering were very limited in earlier years.

Since the invention of LASER in the 1960s, its development of various small-size and low-cost gaseous and solid lasers, and the rapid development of modern electronics and micro-computers within a half-century, especially within the recent three decades, the major difficulties have been overcome. Various new Raman spectrometers and systems have been developed, which made it very easy to perform Raman scattering measurements on different experimental systems and also promote the wide application of Raman

DOI: 10.1201/9781032644912-1

spectroscopy in industry and academic/technological research and applications. Raman laser spectroscopy can be used to diagnose, characterize, and analyze matter and is widely applied in various sciences and engineering, such as physics, chemistry, materials, biology, medicine, environment science, and engineering, in the semiconductor industry and investigation. Raman spectroscopic applications in semiconductors and oxides are rapidly and widely developed. Manuel Cardona has edited a series of books on light scattering in solids, including semiconductors and related materials [1.4–1.6]. There are some more books on Raman scattering on semiconductors and other materials [1.7–1.9]. These books provide useful knowledge and information on Raman scattering investigation on semiconductors and solids in the early years. Recently, Masanobu Yoshikawa contributed a new book, *Advanced Optical Spectroscopy Techniques for Semiconductors: Raman, Infrared, and Cathodoluminescence Spectroscopy Basic Results and Applications* [1.10]. This book focuses on advanced optical spectroscopy techniques, such as Raman, infrared, photoluminescence, and cathodoluminescence spectroscopy. It looks at materials such as silicon and silicon dioxide, nano-diamond thin films, quantum dots, and gallium oxide. It doesn't include SiC, III-nitrides, ZnO, graphene, and more emerging semiconductors and oxides, which are covered in my current book.

In this book, I focus on the Raman scattering of emerging semiconductors (including superlattices and multiple quantum well structures) and oxides, an overview of the research and application results of typical Raman scattering measurements, and analyses of advanced semiconductors and oxides. The studied materials include modern electronic/optoelectronic materials, silicon carbides, III-nitrides, III-V and II-VI compounds, ZnO, GaO, graphene, and other solid materials. These results provide clear evidence that Raman laser spectroscopy is a convenient and powerful tool for the diagnostics of various semiconductors and oxides.

1.2 BASIC THEORIES OF RAMAN SCATTERING

Raman scattering on a semiconductor or solid, that is, inelastic scattering of photons, describes the interaction of laser light with the detected material, resulting in the energy of the laser photons being shifted up or down. The shift in energy gives information about the lattice vibrational modes in the studied material. Raman spectral scan measurements are typically recorded with Raman shifts in wavenumbers, which have units of inverse length, as this value is directly related to energy. To convert between spectral wavelength and wavenumbers of shift in the Raman spectrum, the following formula can be used:

$$\Delta \nu = 1/\lambda_0 - 1/\lambda_1 \tag{1.1}$$

where $\Delta \nu$ is the Raman shift expressed in wavenumber, λ_0 is the excitation wavelength, and λ_1 is the Raman spectrum wavelength. Most commonly, the unit chosen for expressing wavenumber in Raman spectra is inverse centimeters (cm^{-1}). Since wavelength is often expressed in units of nanometers (nm), the formula (1.1) can scale for this unit conversion explicitly, as follows:

$$\Delta \nu (cm^{-1}) = \left[1/\lambda_0 (nm) - 1/\lambda_1 (nm) \right] \times 10^7 (nm)/(cm). \tag{1.2}$$

When a monochromatic light with a frequency v_i is incident into a matter, the interaction of light with matter causes a part of light energy to be absorbed or the matter transfers a portion of energy to the light wave. The scattered or outgoing light will have a new wave frequency v_s. In the concept of quantum mechanics, this scattering process involves two photons, one incident and one scattered, accompanied by the creation or annihilation of a phonon. The conservation of energy and momentum in the scattering process has

$$h v_s = h v_i \pm h v_p \qquad (1.3)$$

$$\mathbf{k}_s = \mathbf{k}_i \pm \mathbf{q}_p \qquad (1.4)$$

where h is a Planck constant; subscripts s, i, and p denote scattered radiation, incident radiation, and phonon; and $\mathbf{k}$ and $\mathbf{q}$ are the wavevectors of the photon and phonon, respectively. The minus sign denotes the emission of a phonon, that is, a Stokes process, and the plus sign denotes the absorption of a phonon or an anti-Stokes process. In semiconductors or solids, the phonon momentum usually has a value of $2\pi/a$, where a is the lattice constant or spacing. The photon momentum ($k = 2\pi/\lambda$) is very small compared to phonon momentum. In Eq. (1.4), the phonon momentum $\mathbf{q}$ is very small also. Therefore, we usually observe the $\mathbf{q} \sim 0$ (near the Γ point of the Brillouin zone of a semiconductor crystal) Raman scattering in the first-order process with one phonon involved. Raman scattering may occur with two or more phonons involved, and in the case of multi-phonon process, the phonon energy item in Eq. (1.1) and the phonon momentum item in Eq. (1.2) should be replaced by the total energy and momentum of all the photons and phonons, respectively. In this case, the $\mathbf{q} \sim 0$ restriction is removed and phonons with large momentum can participate. Eq. (1.5) has a form of a second-order Raman scattering:

$$\mathbf{k}_s - \mathbf{k}_i = \mathbf{q}_p + \mathbf{q}_p' \qquad (1.5)$$

where $\mathbf{q}_{p'}$ is the momentum of a second-order phonon.

A quantum mechanics energy-level representation is used to describe the Raman scattering process, as shown in Figure 1.1. Incident light with a photon energy $h v_i$ interacts with the system, raising the energy to some virtual intermediate state. Relaxation occurs with an energy transfer to (Stokes) or from (anti-Stokes) the system. Simultaneously, a scattered or outgoing light is emitted with the energy $h v_s$.

The intensity ratio between Stokes and anti-Stokes Raman processes depends on the Boltzmann population of the excited energy levels involved and satisfies [1.11, 1.12]:

$$I_{Stokes}/I_{anti-Stokes} = (v_{Stokes}/v_{anti-Stokes})^4 \exp[-h(v_{Stokes} - v_{anti-Stokes})/2kT]$$

$$= [(v_i - \Delta v)/(v_i + \Delta v)]^4 \exp(h\Delta v/kT) \qquad (1.6)$$

where k is the Boltzmann constant, T is temperature, v_{Stokes} is the frequency of the Stokes line, $v_{anti-Stokes}$ is the frequency of the anti-Stokes line, and Δv is the Raman shift. Very

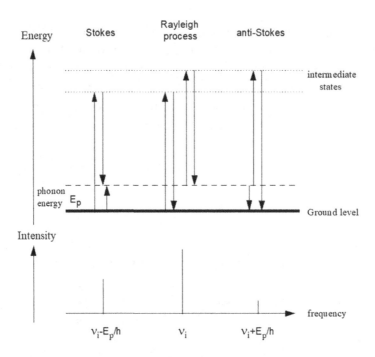

FIGURE 1.1   Representation of Raman scattering process in energy-level diagram.

often, the angular frequency, $\omega = \nu/2\pi$, is used in the Raman spectral analyses. Raman shift is the frequency difference between the incident light frequency and the scattered Stokes light frequency, usually in the unit of wavenumber, cm $^{-1}$.

Raman scattering process can also be described by the Feynman diagram [1.4, 1.13], in which the Raman scattering process is represented graphically for Stokes electron–phonon interaction process with phonon emission, anti-Stokes process with phonon absorption, electron scattering Stokes process, and hole scattering Stokes process. Interested readers can refer to [1.4, 1.13] and references therein for more details.

Raman effect in solids, that is, inelastic scattering of photons by phonons, can be classically treated as the interaction or coupling of the incident electromagnetic field with the phonon field through induced dipole moment. This is accomplished by modifying the electronic polarizability tensor α. Assuming an incident radiation

$$\mathbf{E} = \mathbf{E}_0 \exp(i\alpha t), \tag{1.7}$$

the induced dipole moment is,

$$\mathbf{M} = \alpha \mathbf{E}. \tag{1.8}$$

In Eq. (1.8), the polarizability tensor α with each element α_{kl} may be expanded, according to the normal displacement x, as

$$\alpha_{kl} = \alpha_{0,\,kl} + \alpha_{1,\,kl}x + \alpha_{2,\,kl}x^2 + \dots, \tag{1.9}$$

where $\alpha_{i,kl}$ is the i-th order derivative of α_{kl}. We have

$$\mathbf{M} = \alpha_0 \mathbf{E}_0 \exp(i\omega t) + \alpha_1 \mathbf{x}_0 \mathbf{E}_0 \exp\left(i\left[\omega + \omega_j\right]t\right) + \alpha_2 \mathbf{x}_0{}^2 \mathbf{E}_0 \exp\left(i\left[\omega + 2\omega_j\right]t\right) + \ldots \quad (1.10)$$

The first term describes the Rayleigh scattering process with the radiation frequency unchanged. The second term, which gives the first-order Raman scattering involving the first-order derivative of the polarizability α_1, is generally called the Raman tensor and is denoted by **R** simply.

The scattering cross-section and, consequently, the intensity of Raman scattering are directly determined by the Raman tensor. The values of Raman tensor elements depend on the crystalline structure and the symmetry of the matter. The element semiconductor crystals of Si and Ge (also diamond) possess the diamond structure with the crystalline O_h symmetry in the group theory. The binary compounds of GaAs, InP, InSb, GaSb, InAs, CdTe, ZnSe, and so on, have the zincblende structure with the group T_d symmetry, and some other binary compounds of GaN, InN, AlN, ZnO, and so on have the hexagonal wurtzite structure with the group C_{6v} symmetry. Other types of structures and symmetries may exist for other materials, which are not discussed in this book.

Phonons in solids describe the quantized lattice vibrations, and their frequencies vary with different directions of the Brillouin zone, that is, possess dispersion relations. For a binary compound semiconductor, the longitudinal optical (LO) and transverse optical (TO) phonon branches and longitudinal acoustic (LA) and transverse acoustic (TA) phonon branches are shown with dispersion along the (001) direction in the Brillouin zone. The phonon dispersions of various semiconductors, such as II-VI compounds of CdZnTe and CdSeTe, III-V compounds of InGaSb and AlGaN, and oxides of $PbLaTiO_3$, are presented in the following chapters of this book.

1.3 RAMAN MEASUREMENT CONFIGURATION AND INSTRUMENTS

The electrical constant of a solid is:

$$\varepsilon = (n+ik)^2 \quad (1.11)$$

where n is the refraction index and k is the extinction coefficient. k is related to the material absorption coefficient by the relationship:

$$\alpha = 4\pi k/\lambda \quad (1.12)$$

where λ is the wavelength of the incident light.

Many semiconductors are opaque in the visible or certain wavelength range. Therefore, Raman scattering for these semiconductors is usually performed in the backscattering or quasi-backscattering geometry. Because of the large n values of most semiconductors (much larger than 2), one can still obtain most of the features in the quasi-backscattering as that in the real backscattering. For transparent materials, we can perform Raman measurements in the right-angle geometry arrangement. A large class of semiconductors is opaque in the visible but transparent in the near-infrared (NIR) wavelength range and, therefore, can be performed by Raman measurements in the back- or quasi-backscattering geometry using visible laser excitation but in the right-angle-scattering geometry using a NIR excitation.

The Raman scattering geometry and polarization arrangements can be noted in the following form:

$$Z(X\,X)\overline{Z}, Z(X\,Y)\overline{Z}, Y(X\,X)\overline{Z}, Y(X\,Y)\overline{Z}, \ldots$$

where the first character Z or Y before the parentheses is the direction of the incident laser line, the last character $\overline{z}$ after the parentheses is the scattered light direction, the left character inside the parentheses is the polarization of the incident laser line, and the right character inside the parentheses is the polarization of the scattered light.

In 1928, Sir Raman used sunlight as the source and a telescope as the collector; the detector was his eyes. Thus, such a feeble phenomenon as the Raman scattering was detected remarkably. Early Raman spectroscopy measurements were carried out using the excitation lines from a high press lamp (such as a mercury arc lamp) and with photography recording, which required a long exposure time and made it difficult to detect very weak signals. Since 1960, laser Raman spectroscopy has been fully developed. Many Raman instruments in the 1960s, 70s, and even 80s employed a laser for excitation, a long-length (0.7–1.2 meter or longer) double or triple grating spectrometer for dispersion, a photomultiplier tube (PMT) for optical signal detection and conversion into electronic signals, lock-in amplifier or photon counting electronics for signal amplification, and recorder or computer for data display and treatments. These types of Raman instruments, still existing in many universities and laboratories, have been used to produce many nice works on semiconductors and microstructures. In brief, they employ the single-channel detection technique, monitor one wavelength at a time by scanning the spectrometer, and record the Raman signals at this wavelength (wavenumber). After the spectrometer finishes the data recording for one wavenumber, it moves to another wavenumber and records again. However, this process causes a considerable waste of time to obtain a spectrum over a certain wavelength range with an acceptable signal-to-noise ratio. Spontaneous Raman scattering is typically very weak. For many years, Raman spectrometers used holographic gratings with multiple dispersion stages to achieve a high degree of laser rejection and photomultipliers as the detectors. In the mid-1980s, the student partner, Angelo Mascarenhas, and I, under the supervision of Prof. W. J. Choky, built up the first Raman system using a SPEX 1401 spectrometer with a PMT as the detector, at long acquisition times for detection, but produced nice papers of Raman scattering on 3C–SiC/Si materials [1.13, 1.14]. However, modern instrumentation almost universally employs notch or edge filters for laser rejection. Dispersive single-stage spectrographs (axial transmissive or Czerny–Turner monochromators) paired with charge-coupled device (CCD) detectors are most common.

Since the late 1970s, a new generation of Raman instruments using optical multichannel analyzers (OMA) with Si detector array or CCD has been developed, which collects a certain wavelength range of light at the same time and accumulates the signals simultaneously, leading to a much-improved spectral signal-to-noise ratio over a certain spectral range and with a short experimental time. In the 1970s and 1980s, due to the high prices,

OMA systems and detectors were limited in application. Since the late 1980s, the prices have dropped considerably, and sensitivity and performance quality on weak signal detection have been greatly improved. Three stages of grating were used to depress the laser tail signals in the low wavenumber range, close to the laser excitation line, from Raman measurements. Prof. Sidney Perkowitz and collaborators and I have employed a commercial SPIE-1877 system with OMA to work on many II-VI and III-V semiconductor materials in 1988–92, extension to 1995, with more outputs reported in the next chapters.

Later, an advanced type of notch filter was developed. It cuts off the light from a particularly designed wavelength (laser line) and can be put near the entrance slit of the spectrometer to depress the laser tail signals greatly. By using the notch filter with careful optical design and alignment, the Raman instrument comes back to a single grating spectrometer without using two or three gratings. Because of the strong optical signal pass-through ability, it can be easily connected with a microscope to develop images, two-dimensional mapping, and three-dimensional display. Since 1992 up to now, more than three decades, many collaborators and I have used these types of Raman microscopies, including those manufactured by Renishaw [1.15], Horiba [1.16], Thermo Fisher [1.17], and so on, to have produced many scientific outputs, to be introduced in following chapters. For ordinary characterization of semiconductors on mainly the optical phonons, it is quite good. Its advantages of fast and convenient operation make it a very powerful nondestructive characterization tool. Figure 1.2 shows the block sketch of this type of new Raman instrument. More details can be found from K. P. J. Williams et al. [1.18].

This type of Raman instrument with notch filter–single spectrograph–CCD has two major limits: (a) it permits to approach 100 or 50 cm^{-1} of the laser line, depending on the property of the notch filter; and (b) it permits the minimum data interval of about 1 cm^{-1} currently. Therefore, in the study of acoustic phonon modes in most semiconductors and the folded modes in semiconductor superlattices in the range far below 100 cm^{-1} or performing the high-resolution (better than 1 cm^{-1}) Raman measurements, we must employ the scanning double or triple spectrometer Raman systems; for example, we could use a

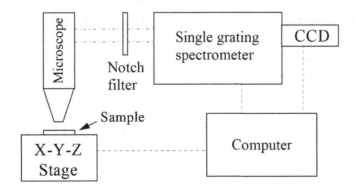

FIGURE 1.2 Block sketch of a micro-Raman system using a microscope–notch filter–single grating spectrometer–CCD detector.

more advanced T64000 triple monochromatic stage Raman spectrometer, although more expensive [1.19].

1.4 RAMAN TENSOR AND SELECTION RULE

The differential cross-section per unit volume or the Raman scattering efficiency, S, for the cubic (diamond and zincblende) and wurtzite crystal can be written as [1.20]:

$$S = (\omega_s / c)^4 (N\hbar / 2M\omega)(n_0 + 1)| \sum_j \mathbf{e}_s \cdot R_j \cdot \mathbf{e}_i |^2, \qquad (1.13)$$

where ω_s is the frequency of the scattered light, c is the velocity of light in vacuum, N is the number of primitive cells per unit volume, M is the reduced mass of the vibrating atoms, ω_s is the phonon frequency, and n_0 is the Bose–Einstein population factor. $\mathbf{e}_i$ and $\mathbf{e}_s$ are the dimensionless vectors describing the polarization of the incident and scattered light, respectively. R_j represents the Raman tensor of the phonon j.

The zincblende cubic crystal with the T_d symmetry has Raman-active modes associated with the singly degenerated A_1, doubly degenerated E, and triply degenerated T_2 irreducible representations. Using x, y, and z along the crystal axes of (100), (010), and (001), respectively, their Raman tensors have the form [1.21]:

$$A_1 : \begin{bmatrix} a & 0 & 0 \\ 0 & a & 0 \\ 0 & 0 & a \end{bmatrix}, E : \begin{bmatrix} b/2 & 0 & 0 \\ 0 & -b/2 & 0 \\ 0 & 0 & b \end{bmatrix}, \begin{bmatrix} \sqrt{3}b/2 & 0 & 0 \\ 0 & -\sqrt{3}b/2 & 0 \\ 0 & 0 & 0 \end{bmatrix}, \text{ and}$$

$$T_2 : \begin{bmatrix} 0 & 0 & 0 \\ 0 & 0 & c \\ 0 & c & 0 \end{bmatrix}, \begin{bmatrix} 0 & 0 & c \\ 0 & 0 & 0 \\ c & 0 & 0 \end{bmatrix} \text{ and } \begin{bmatrix} 0 & c & 0 \\ c & 0 & 0 \\ 0 & 0 & 0 \end{bmatrix}. \qquad (1.14)$$

We can see the Raman spectral line intensity will be dependent on the measurement conditions of the incident/collect direction, the polarization arrangement, and the crystalline orientation of the sample. Because of the features of these Raman tensors, some commonly used selection rules can be obtained for cubic structural materials, listed in Table 1.1.

The C_{6v} symmetric wurtzite crystal has Raman-active modes of A_1, E_1, and E_2 with the following Raman tensors:

$$A_1(z) = \begin{bmatrix} a & 0 & 0 \\ 0 & a & 0 \\ 0 & 0 & b \end{bmatrix}, E_1(x) = \begin{bmatrix} 0 & 0 & c \\ 0 & 0 & 0 \\ c & 0 & 0 \end{bmatrix}, E_1(y) = \begin{bmatrix} 0 & 0 & 0 \\ 0 & 0 & c \\ 0 & c & 0 \end{bmatrix}, \text{ and}$$

$$E_2 = \begin{bmatrix} 0 & d & 0 \\ d & 0 & 0 \\ 0 & 0 & 0 \end{bmatrix}, \begin{bmatrix} d & 0 & 0 \\ 0 & -d & 0 \\ 0 & 0 & 0 \end{bmatrix}. \qquad (1.15)$$

TABLE 1.1   Some Frequently Used Raman Selection Rules for Cubic Crystals

Surface Orientation	Polarization Arrangement	Selection Rule	
		TO	LO
<001>	$Z(XX)\bar{Z}$	Forbidden	Forbidden
	$Z(XY)\bar{Z}$	Forbidden	Allowed
	$Z(X'X')\bar{Z}$	Forbidden	Allowed
	$Z(X'Y')\bar{Z}$	Forbidden	Allowed
	Z:<001>, X:<100>, Y:<010>; X':<1$\bar{1}$0>, Y'<110>		
<011>	$Z(XX)\bar{Z}$	Allowed	Forbidden
	$Z(XY)\bar{Z}$	Allowed	Forbidden
	Z:<011>, X:<100>, Y:<01$\bar{1}$>		
<111>	$Z(XX)\bar{Z}$	Allowed	Allowed
	$Z(XY)\bar{Z}$	Allowed	Allowed
	Z:<111>, X:<$\bar{2}$11>, Y:<01$\bar{1}$>		

TABLE 1.2   Some Frequently Used Raman Selection Rules for Wurtzite C_{6v} Crystals

	Scattering Configuration	Allowed Modes
Backscattering	$Z(XX)\bar{Z}$	E_2 and $A_1(LO)$
	$Z(XY)\bar{Z}$	E_2
	$Y(XX)\bar{Y}$	E_2 and $A_1(TO)$
	$Y(ZZ)\bar{Y}$	$A_1(TO)$
	$Y(ZX)\bar{Y}$	$E_1(TO)$
Right-angle-scattering	$X(ZX)Y$	$E_1(TO)$ and $E_1(LO)$
	$X(ZZ)Y$	$A_1(TO)$
	$X(YX)Y$	$E_1(TO)$ and $E_1(LO)$

Some frequently used selection rules [1.22] are listed in Table 1.2, where we set the crystal c-axis along the Z:<001> direction.

The polar mode $A_1(z)$ is polarized along the z-axis of the crystal, and the polar $E_1(x, y)$ modes are polarized in the xy-plane. The polarity of the modes and the uniaxial structure of the hexagonal crystal led to frequency dispersion of the polar modes, depending on the propagation direction. The $A_1(z)$ phonon propagating along the z-axis is purely LO with the highest frequency due to the contribution of the electric field, and its mode propagating in the xy-plane is purely TO with the lowest frequency. $E_1(x, y)$ phonons are purely transverse for propagation along the z-axis and transverse or longitudinal in the xy-plane [1.23].

1.5 OPTICAL PHONON DATA OF MAJOR ELEMENT AND BINARY SEMICONDUCTORS

We list below some data, including the lattice constants, energy band gap (direct or indirect), and optical phonon TO and LO (at the center Γ point of the Brillouin zone)

frequencies, of main element and binary semiconductors with diamond, zincblende, and wurtzite structures, in Tables 1.3 and 1.4, respectively. These cited data are mainly from [1.24, 1.25], data for (Al,Ga,In)N are from [1.26, 1.27] and for SiC from [1.28], with some from our data and other references mentioned in the next chapters.

TABLE 1.3   Data of Element and Binary Semiconductors with Diamond and Zincblende Structures (300 K)

Semiconductor	Lattice Constant (Å)	Energy band gap		Phonon Frequency	
		Direct (eV)	Indirect (eV)	TO (Γ) (cm^{-1})	LO (Γ) (cm^{-1})
Diamond	3.567		5.5	1333	1333
Si	5.431		1.124	520	520
Ge	5.658		0.805	300.6	300.6
3C–SiC	4.3596		2.416 (2 K)	796.2	972.7
AlAs	5.66	3.03	2.153	361.7	403.7
AlP	5.4635	3.62	2.505 (4 K)	439.4	501
AlSb	6.1355	2.3	1.615	318.7	340
GaAs	5.6532	1.424		269	292
GaN (cubic)	4.503			555	740
GaP	5.4505	2.78	2.272	365.3	402.3
GaSb	6.0959	0.75		223.6	232.6
InAs	6.0583	1.3543		217.3	238.6
InP	5.8687	1.344		303.3	344.5
InSb	6.4794	0.18		179.7	190.7
CdS (cubic)	5.818	2.554		234.7 (77 K)	302.2 (77 K)
CdSe	6.052	1.751		169	209
CdTe	6.482	1.475		140	169
HgTe (cubic)	6.456	−0.141		116	139
ZnS (cubic)	5.4102	3.68		273	350
ZnSe	5.6676	2.7		205	250
ZnTe	6.1037	2.3941 (1.6 K)		177	205

TABLE 1.4   Data of Binary Semiconductors with Wurtzite Structures (300 K), Including the Corresponding Lattice Constant a and c (Å), Energy Gap E_g (eV), and Phonon Frequency (cm^{-1})

Semiconductor	a (Å)	c (eV)	E_g (eV)	E$_2$ (low) E$_2$ (high) Phonon Frequency (cm^{-1})		A$_1$(TO) E$_1$(TO) A$_1$(LO) E$_1$(LO)				
AlN	3.111	4.978	6.2		248	657	610	668	886	911
GaN	3.189	5.185	3.40	141	568	538	557	733	742	
InN	3.544	5.718	0.65	88	490	495	460	590	596	
ZnO	3.253	5.213	3.28	98	438	378	410	576	588	
6H–SiC	3.081	15.117	2.86	150	767	788	797	965		
4H–SiC	3.080	10.081	3.20	204	776.5	778	796			
							964			
							998			

1.6 RAMAN SPECTRAL LINE SHAPE ANALYSES AND SPATIAL CORRELATION MODEL

Raman spectral line shape analysis can provide useful information such as peak frequency, width, intensity, and their dependence on crystal quality. The basic and simple line shape analysis adapts Lorentz and Gauss functions as well as the Voigt function, a convolution of Gaussian and Lorentzian [1.29, 1.30]. In an ideal semiconductor, the crystal lattice translation symmetry leads to plane wave phonon eigenstates. Due to the energy and momentum conservation, only q = 0 phonons at the center of the Brillouin zone (Γ point) participate in the first-order Raman scattering process. However, disorders or finite-size effects may partially or completely relax the momentum conservation, leading to a downshift and broadening of the Raman peak. A quantitative model was presented by Richter et al. [1.31] and developed by Parayanthal and Pollak [1.32] and named as spatial correlation model (SCM).

The TO mode is a critical factor for the integrity of Raman spectra. Using the SCM, the first-order TO Raman scattering intensity I (ω) is given by [1.32, 1.33]:

$$I(\omega) \propto \int_0^1 \exp\left(-q^2 L^2 \Big/ 4\right) \frac{d^3 q}{\left[\omega - \omega(q)\right]^2 + \left(\Gamma_0 / 2\right)^2} \tag{1.16}$$

where q is expressed in units of $2\pi/a$, a is the lattice constant, and L is the correlation length in units of a, representing the phonon propagation length and characterizing the crystalline perfection of the material. The dispersion relation for optical phonons can be represented by an analytical form:

$$\omega^2(q) = A + \{A^2 - B[1 - con(\pi q)]\}^{1/2}, \tag{1.17}$$

or

$$\omega(q) = A - Bq^2 \tag{1.18}$$

where A and B are adjustable parameters [1.32, 1.33]. This SCM has been employed by the author and collaborators to analyze many semiconductors and oxides, such as SiC, GaN, AlN, ZnO, GaO, and graphene, reported in the next chapters.

1.7 RAMAN DETERMINATION OF CARRIER CONCENTRATION AND STRESS

Raman scattering can offer a nondestructive experimental technique for the determination of the carrier concentration due to doping in semiconductors through the LO phonon and plasma coupling (LOPC). There exist three mechanisms responsible for the coupling of LO phonon and plasmon, deformation-potential, electro-optic, and density-fluctuation mechanism. In wide-bandgap semiconductors such as SiC, GaN, and GaP, the deformation-potential mechanism and the electro-optic mechanism are dominant compared with the density-fluctuation mechanism. A set of formulas on LOPC was developed to determine the free-carrier concentration in these wide-bandgap semiconductors. The Raman intensity of LOPC mode can be expressed as [1.34, 1.35]:

$$I_{LOPC} = \frac{d^2S}{d\omega d\Omega}\bigg|_A = \frac{16\pi h n_2}{V_0^2 n_1} \frac{\omega_2^4}{C^4} \left(\frac{d\alpha}{dE}\right)(n_\infty + 1) A \text{Im}\left(-\frac{1}{\varepsilon}\right) \tag{1.19}$$

$$A = 1 + 2C\frac{\omega_T^2}{\Delta}\left[\omega_p^2\gamma\left(\omega_T^2 - \omega^2\right) - \omega^2\eta\left(\omega^2 + \gamma^2 - \omega_p^2\right)\right] + C^2\left(\frac{\omega_T^4}{\Delta\left(\omega_L^2 - \omega_T^2\right)}\right.$$

where

$$\left.\times\left\{\omega_p^2\left[\gamma\left(\omega_L^2 - \omega_T^2\right) + \eta\left(\omega_p^2 - 2\omega^2\right)\right] + \omega^2\eta\left(\omega^2 + \gamma^2\right)\right\}\right) \tag{1.20}$$

$$\Delta = \omega_p^2\gamma\left[\left(\omega_T^2 - \omega^2\right)^2 + (\omega\eta)^2\right] + \omega^2\eta\left(\omega_L^2 - \omega_T^2\right)\left(\omega^2 + \gamma^2\right) \tag{1.21}$$

and ω_L is the LO mode frequency; ω_T is TO mode frequency; η is phonon damping constant; γ is plasma damping constant; n_1 and n_2 are refractive indices at incident frequency ω_1 and scattering frequency ω_2, respectively; C is Faust-Henry coefficient, here the value is about 0.35; α is polarizability; E is macroscopic electric field; and n_ω is Bose–Einstein factor. The dielectric function is described as:

$$\varepsilon = \varepsilon_\infty\left(1 + \frac{\omega_L^2 - \omega_T^2}{\omega_T^2 - \omega^2 - i\omega\eta} - \frac{\omega_p^2}{\omega(\omega + i\gamma)}\right) \tag{1.22}$$

$$\omega_p^2 = \frac{4\pi n e^2}{\varepsilon_\infty m^*} \tag{1.23}$$

where ω_p is the plasma frequency, n is free-carrier concentration, m^* is effective mass while e is the unit charge, and ε_∞ is the high-frequency dielectric constant. For polar semiconductors, strong coupling between the LO phonon and the free-carrier plasmon exists. By taking the above C and other fitting parameters, the line shape of the LO phonon–plasmon coupled mode can be fitted by means of the least-square difference. This method has been successfully applied to obtain the carrier concentrations in GaP [1.36], CdTe [1.37], 6H–SiC [1.34], 3C–SiC [1.35], 4H–SiC [1.34], GaN [1.33], and so on. More from Zhe Chuan Feng and collaborators will be presented in the following chapters.

Raman spectral line frequencies are sensitive to the stress and strain in bulk and film semiconductors. These Raman shifts with strain and stress have been used for the measurements of micro-strains and stresses. Epitaxial growth of thin layer films on a thick substrate may lead to layer stresses and strains due to the differences in lattice constants and thermal coefficients between the substrate and the film. The layer stress component in the direction perpendicular to the surface of the substrate is different from that parallel to the surface of the substrate. The components of stress and strain in the parallel plane can be regarded as isotropic. Raman measurements are sensitive, precise, nondestructive, and convenient. They are applied to the determination of biaxial stress in thin semiconductor films grown on thick substrates such as cubic SiC on Si [1.38], in which we described a general relation of Raman phonon frequencies with strain and stresses in the case of a generalized axial stress distribution. The hydrostatic pressure, uniaxial, and biaxial stresses are only special cases of this generalized stress. A series of CVD-grown 3C–SiC/Si(100) samples with SiC film thicknesses from 600 Å to 17 μm have been measured with Raman spectra [1.14], and stresses and strains in those films of 4–17 μm thick were calculated [1.38]. The stresses in the 3C–SiC film are in the order of 10^9 dyn/cm^2 or 10^8 Pa, and the strains are 0.1–0.2%, despite a high lattice mismatch of ~20% between bulk 3C–SiC and Si. Raman E_2 mode peak shift is used to determine the stress and strain in the III-nitride films, such as GaN and AlN grown on sapphire or other lattice mismatched substrates, to be discussed in later Chapters 3 and 4.

1.8 TEMPERATURE DEPENDENCE OF RAMAN SCATTERING

A model involving three- and four-phonon processes is used to fit Raman shift and width. The temperature dependence of Raman frequency $\omega(T)$ can be modeled by Eqs. [1.39, 1.40]:

$$\omega(T) = \omega_0 + \omega^{(1)}(T) + \omega^{(2)}(T) \tag{1.24}$$

where ω_0 is the harmonic frequency of the optical mode phonon at a temperature near absolute zero; $\omega^{(1)}(T)$ is the term caused by linear thermal expansion, and $\omega^{(2)}(T)$ denotes anharmonic phonon coupling contribution to Raman shift. The second term $\omega^{(1)}(T)$ can be expressed as

$$\omega^{(1)}(T) = \omega_0 \left\{ \exp\left[-\gamma \int_0^T (\alpha_c(t) + 2\alpha_a(t)) dt \right] - 1 \right\}, \tag{1.25}$$

where γ is the Grüneisen parameter for optical Raman mode, $\alpha_c(t)$ and $\alpha_a(t)$ are coefficients of linear thermal expansion along the direction of c and a axis. The third term $\omega^{(2)}(T)$ can be written as:

$$\omega^{(2)}(T) = M_1 \left\{ 1 + \sum_{i=1}^2 \frac{1}{e^{x_i} - 1} \right\} + M_2 \left\{ 1 + \sum_{j=1}^3 \left[\frac{1}{e^{y_j} - 1} + \frac{1}{(e^{y_j} - 1)^2} \right] \right\}, \tag{1.26}$$

where M_1 and M_2 are fitting parameters. The value of the exponent x_i and y_j are given as $\Sigma x_i = \Sigma y_j = \hbar\omega_0$. The first term of $\omega^{(2)}(T)$ describes the three-phonon process, and the second denotes the four-phonon process.

Raman linewidth (Γ) of phonon scattering peaks can be described in terms of anharmonic interactions. Temperature dependence of Raman linewidth can be expressed as [1.39, 1.40]:

$$\Gamma(T) = \Gamma_0 + N_1 \left\{ 1 + \sum_{i=1}^2 \frac{1}{e^{x_i} - 1} \right\} + N_2 \left\{ 1 + \sum_{j=1}^3 \left[\frac{1}{e^{y_j} - 1} + \frac{1}{(e^{y_j} - 1)^2} \right] \right\}, \tag{1.27}$$

where Γ_0 is the Raman linewidth at 0 K which results from the scattering of impurity and defects. Quantities N_1 and N_2 are fitting parameters. Exponents x_i and y_j are given as $x_1 = x_2 = \hbar\omega_0/2$ and $y_1 = y_2 = y_3 = \hbar\omega_0/3$. The second and third terms of $\Gamma(T)$ represent three- and four-phonon processes, respectively. Eqs. (1.24–1.27) have been successfully used by our research teams to analyze the temperature-dependent Raman spectra (TDRS) for GaN [1.39] and 4H–SiC [1.40], which are shown in detail in the following chapters.

Also, an empirical formula was proposed to describe the temperature dependence of the Raman line position, by J. B. Cui et al. [1.41]:

$$\omega(T) = \omega_0 - A/\{\exp[B(\hbar\omega_0/kT)] - 1\} \tag{1.28}$$

where ω_0 is the Raman phonon frequency at 0 K, and ω_0, A, and B are the fitting parameters. The full width at half maximum (FWHM) values can be fitted by the following formula, proposed by R. Han and others [1.42]:

$$\Gamma(T) = \Gamma_0 + C\left(1 + \frac{2}{e^{h\omega_0/kT} - 1}\right) + D\left(1 + \frac{3}{e^{h\omega_0/3kT} - 1} + \frac{3}{\left(e^{h\omega_0/3kT} - 1\right)^2}\right) \tag{1.29}$$

where Γ_0 is the Raman FWHM at 0 K, and Γ_0, C, and D are fitting parameters. The second term describes the asymmetric decay of the three-phonon process, and the third term describes the symmetric decay of the four-phonon process. The Raman line width FWHM is related to the phonon lifetime τ, following [1.43]:

$$\frac{\Delta E}{\hbar} = \frac{1}{\tau} \tag{1.30}$$

Eqs. (1.28–1.30) have been used to analyze the TDRS of 3C–SiC grown on 4H–SiC by us [1.43], which is further discussed in Chapter 2 of this book.

1.9 RAMAN APPLICATIONS IN SOLIDS, SEMICONDUCTORS, AND OXIDES

Since 1992 to now, I have edited and published 12 review-style books (with 2 co-edited) [1.27–1.29, 1.44–1.51]. In the past century, two books [1.44, 1.45] focused on old II-VI, III-V, and SiC semiconductors, and one [1.46] was for porous silicon. In the 21st century, I have mainly concerned wide-bandgap semiconductors of SiC [1.29, 1.47, 1.48], III-nitrides, nanoengineering [1.49], and ZnO materials and devices [1.51]. Among the above, Raman scattering was used as one of the characterization and investigation tools. The current book emphasizes the Raman scattering for a wide range of emerging semiconductors and oxides and especially covers Raman application on ultra-wide-bandgap semiconductors of AlN, Ga_2O_3, and graphene, as well as ferroelectric oxides and others.

Raman spectroscopy is used in chemistry to identify molecules and study chemical bonding and intramolecular bonds. Because vibrational frequencies are specific to a molecule's chemical bonds and symmetry (the fingerprint region of organic molecules is in the wavenumber range of 500–1,500 cm^{-1}), Raman provides a fingerprint to identify molecules. For instance, Raman and infrared spectra were used to determine the vibrational frequencies of SiO, Si$_2$O$_2$, and Si$_3$O$_3$ based on normal coordinate analyses.

In solid-state physics, Raman spectroscopy is used to characterize materials, measure temperature, and find the crystallographic orientation of a sample. As with single molecules, a solid material can be identified by characteristic phonon modes. Information on the population of a phonon mode is given by the ratio of the Stokes and anti-Stokes intensity of the spontaneous Raman signal. Raman spectroscopy can also be used to observe other low-frequency excitations of a solid, such as plasmons, magnons, and superconducting gap excitations. Distributed temperature sensing uses the Raman-shifted backscatter from laser pulses to determine the temperature along optical fibers. The orientation of an anisotropic crystal can be found from the polarization of Raman-scattered light with respect to the crystal and the polarization of the laser light, if the crystal structure's point group is known.

The Raman effect is based on the inelastic scattering process between the incident photons and the phonons in materials. When the photons of incident light interact with the vibration phonons of the materials, the light is scattered at a frequency and the difference in frequency between the incident and the scattered light provides the information of the lattice vibrations. Raman vibration spectra are widely used for providing a structural fingerprint for molecule identification.

Raman vibration spectra of materials are significantly influenced by microstructural changes, impurities, residual stress, and so on, which leads to changes in phonon frequency, breakdown of Raman selection rules, and other effects. These defects and subsurface damages induced by crystal growth and micro/nano-machining can be characterized using Raman information, including band position, band position shift, FWHM, and intensity. Raman spectroscopy technique is an efficient, powerful, sample-preparation-friendly, and nondestructive testing method to characterize these defects and subsurface damages.

Raman spectroscopy technique has been tremendously improved to overcome problems like fluorescence, poor sensitivity, or weak Raman signals. In addition, compared with the conventional dispersive Raman technique, many advanced Raman techniques have been developed to meet the demands of analysis. Confocal Raman microscopy can provide three-dimensional images of material composition with micrometer resolution and clear image quality. Resonant Raman scattering (RRS) allows researchers to explore the material spectra in the range of energies of the photon energy itself (typically 1–3 eV). This book aims to present the principles and applications of the Raman spectroscopy technique in micro/nano-machining. The theory of Raman spectroscopy is introduced, including the principle of Raman spectroscopy, the relationship between incident light wavelength and penetration depth, laser spot size, and spectral resolution. Typical Raman spectroscopy analysis in the field is introduced, including stacking faults, phase transformation, and residual stress characterization.

The Raman effect is understood usually as the scattering of light due to phonons; any kind of inelastic light scattering is called Raman scattering. The scattering by electronic excitations or magnetic excitations is also Raman scattering, where the coupling is via virtual electronic states. Electron–phonon interaction is actually very important within the electronic transitions itself since it renormalizes the electronic gap of the materials and gives a broadening to the electronic states. When the electrons have enough energy to emit optical phonons, the photoluminescence processes become more efficient since the emission probability increases.

In a semiconductor, most of the incident radiation is reflected, absorbed, or transmitted; only a small amount is scattered. The elastically scattered light is called Rayleigh scattering. Elastic scattering is possible due to the existence of defects in a sample; in a nearly perfectly grown material with translational symmetry, the elastic scattering must be very low. There exist some special features to be concerned about, as below:

- In nonpolar semiconductors (Si, diamond), no splitting occurs between the transverse (TO) and longitudinal (LO) optical phonons, but in polar semiconductors (GaAs, GaN, and ZnO), a splitting related to the difference between the low- and high-frequency dielectric permittivity (effective charges) and LO–TO splitting occurs.

- Splitting is a measure of the polarity of semiconductors. In polar semiconductors, plasmons interact with LO phonons, giving rise to an intercrossing when the plasma frequency coincides with the LO phonon frequency, depending on the electron concentration. From Raman spectral analyses, electron concentration can be obtained.

- Since phonons depend on the mass of the ions, Raman scattering has been used to study the isotopic composition of isotopically engineered samples and phonon anharmonicities.

- Phonons are also sensitive to pressure or stress variations.

- In thin films and semiconductor nanostructures, Raman scattering is used very often to obtain the lattice mismatch through the phonon deformation potentials.

- Exciton plays an important role in the first-order RRS or Raman scattering by one phonon. In the second-order RRS, exciton also plays a fundamental role in the resonant behavior. Electron–phonon interaction is a crucial part of the resonant process.

- Electron–phonon interaction is very important within the electronic transitions which give a broadening to the electronic states. It is important not only in optics but also in transport, through the different relaxation mechanisms. Also, when electrons have enough energy to emit optical phonons, photoluminescence processes become more efficient since the emission probability increases.

- In nonpolar semiconductors, there is no splitting between the optical transverse (TO) and longitudinal (LO) phonons, but in polar semiconductors like GaAs, there is a splitting related to the difference between the low- and high-frequency dielectric permittivity (or the effective charges). The splitting is a measure of the polarity of the semiconductor. Plasmons interact with LO phonons, giving rise to an LO plasma coupling (LOPC) mode.

- From the measurement of two split frequencies, the electron concentration can be obtained. Electrons interact strongly with the phonons, giving rise to the Fano line shape, produced by the sum of quantum probabilities of the two processes.

REFERENCES

[1.1] C. V. Raman and K. S. Krishnan, "The negative absorption of radiation", Nature **122**, 12–13 (1928). https://doi.org/10.1038/122012b0.

[1.2] A. Smekal, "Zur Quantentheorie der Dispersion (German, i.e., Dispersion Quantum Theory)", Naturwissenschaften **11**, 873–875 (1923). https://doi.org/10.1007/BF01576902.

[1.3] G. S. Landsberg and L. I. Mandelstam, "New phenomenon in the scattering of light (preliminary report)", J. Russ. Phys. Chem. Soc. Phys. Sec. **60**, 335 (1928).

[1.4] M. Cardona (Ed.), Light Scattering in Solids I, Springer, Berlin (1975).

[1.5] M. Cardona (Ed.), Light Scattering in Solids V: Superlattices and Other Microstructures (Topics in Applied Physics, 66), Springer, Berlin (1989).

[1.6] M. Cardona and R. Merlin (Eds.), Light Scattering in Solids IX: Novel Materials and Techniques (Topics in Applied Physics, 108), Springer, Berlin (2007).

[1.7] T. Ruf, Phonon Raman Scattering in Semiconductors, Quantum Wells and Superlattices: Basic Results and Applications, Springer, Berlin (1998).

[1.8] W. H. Weber and R. Merlin (Eds.), Raman Scattering in Materials Science, Springer, Berlin (2000).

[1.9] I. R. Lewis and H. G. M. Edwards (Eds.), Handbook of Raman Spectroscopy. From the Research Laboratory to the Process Line, Springer, Berlin (2001).

[1.10] M. Yoshikawa, Advanced Optical Spectroscopy Techniques for Semiconductors: Raman, Infrared, and Cathodoluminescence Spectroscopy Basic Results and Applications, Springer, Berlin (2023).

[1.11] T. Goldstein, S.-Y. Chen, J. Tong, D. Xiao, A. Ramasubramaniam, and J. Yan, "Raman scattering and anomalous stokes – anti-stokes ratio in $MoTe_2$ atomic layers", Sci. Rep. **6**, 28024 (2016). https://doi.org/10.1038/srep28024.

[1.12] K. Thapliyal and J. Peřina Jr., "Ideal pairing of the stokes and anti-stokes photons in the Raman process", Phys. Rev. A **103**, 033708 (2021). https://doi.org/10.1103/PhysRevA.103.033708.

[1.13] R. D. Mattuck, A Guide to Feynman Diagrams in the Many-Body Problem, Dover Publications, New York (1992).

[1.14] Z. C. Feng, A. J. Mascarenhas, W. J. Choyke, and J. A. Powell, "Raman scattering studies for chemical vapor deposited 3C-SiC films on (100) Si", J. Appl. Phys. **64**, 3176–3186 (1988). https://doi.org/10.1063/1.341533.

[1.15] Renishaw Raman spectroscopy. www.renishaw.com/en/raman-spectroscopy-6150.

[1.16] Horiba Raman spectroscopy. www.horiba.com/us/en/scientific, www.ramanacademy.com.

[1.17] Thermo Fisher Scientific Raman system. www.thermofisher.com/us/en/home/industrial/spectroscopy-elemental-isotope-analysis/molecular-spectroscopy/raman-

[1.18] K. P. J. Williams, G. D. Pitt, D. N. Batchelder, and B. J. Kip, "Confocal Raman microspectroscopy using a stigmatic spectrograph and CCD detector", Appl. Spectrosc. **48**, 232–235 (1994). https://doi.org/10.1366/0003702944028407.

[1.19] Horiba Jobin Yvon T64000 Raman spectrometer. www.horiba.com/fileadmin/uploads/Scientific/Documents/Raman/HSC-T64000_U1000-2013-V1.pdf

[1.20] J. M. Calleja, H. Vogt, and M. Cardona, "Absolute Raman scattering efficiencies of some zincblende and fluorite-type materials", Philos. Mag. A **45**, 239–254 (1982). https://doi.org/10.1080/01418618208244297.

[1.21] R. Loudon, "The Raman effect in crystals", Adv. Phys. **50**, 813–864 (2001). https://doi.org/10.1080/00018730110101395.

[1.22] T. Azuhata, T. Sota, K. Suzuki, and S. Nakamura, "Polarized Raman spectra in GaN", J. Phys. Condens. Matter **7**, L129 (1995). https://doi.org/10.1088/0953-8984/7/10/002.

[1.23] D. Kirillov, H. Lee, and J. S. Harris, Jr., "Raman scattering study of GaN films", J. Appl. Phys. **80**, 4058 (1996). https://doi.org/10.1063/1.363367.

[1.24] O. Madelung (Ed.), Landolt-Bornstein Numerical Data and Functional Relationships in Science and Technology, New Series, Vol. 22a, IntFinsical Properties of Group IV Elements and III-V, II-M and I-WI Compounds, Springer-Verlag, Berlin (1987).

[1.25] S. Adachi, Optical Properties of Group-IV, III-V and II-VI Semiconductors, John Wiley & Sons, Chichester (2005). https://doi.org/10.1002/0470090340.

[1.26] Z. C. Feng (Ed.), Handbook of Solid-State Lighting and LEDs (24-chapters), CRC Press, Taylor & Francis Group, London; New York (2017). https://doi.org/10.1201/9781315151595.

[1.27] Z. C. Feng (Ed.), III-Nitride Materials, Devices and Nanostructures, World Scientific Publishing Press, Singapore, p. 410 (2017). https://doi.org/10.1142/q0092.

[1.28] Z. C. Feng (Ed.), Handbook of Silicon Carbide and Related Materials (15-chapters), CRC Press, Taylor & Francis Group, London; New York (2023). www.routledge.com/9780367188269.

[1.29] V. Jain, M. C. Biesinger, and M. R. Linford, "The gaussian-lorentzian sum, product, and convolution (voigt) functions in the context of peak fitting X-ray photoelectron spectroscopy (XPS) narrow scans", Appl. Surf. Sci. **447**, 548 (2018). https://doi.org/10.1016/j.apsusc.2018.03.190.

[1.30] Spectral line shape – Wikipedia. https://en.wikipedia.org/wiki/Spectral_line_shape.

[1.31] H. Richter, Z. P. Wang, and L. Ley, "The one phonon Raman spectrum in microcrystalline silicon", Solid State Commun. **39**, 625–629 (1981). https://doi.org/10.1016/0038-1098(81)90337-9.

[1.32] P. Parayanthal and F. H. Pollak, "Raman scattering in alloy semiconductors", Phys. Rev. Lett. **52**, 1822–1825 (1984). https://doi.org/10.1103/PhysRevLett.52.1822.

[1.33] Z. C. Feng, "Micro-Raman scattering and micro-photoluminescence of GaN thin films grown on sapphire by metalorganic chemical vapor deposition", Opt. Eng. **41**, 2022–2031 (2002). https://doi.org/10.1117/1.1489051.

[1.34] H. Harima, S. Nakashima, and T. Uemura, "Raman scattering from anisotropic LO-phonon – plasmon – coupled mode in n-type 4H- and 6H-SiC", J. Appl. Phys. **78**, 1996–2005 (1995). https://doi.org/10.1063/1.360174.

[1.35] S. Chen, X. Jiang, C.-C. Tin, L. Wan, and Z. C. Feng, "Adducing crystalline features from Raman scattering studies of cubic SiC using different excitation wavelengths", J. Phys. D Appl. Phys. **50**, 115102 (2017). https://doi.org/10.1088/1361-6463/aa5626.

[1.36] G. Irmer, V. V. Toporov, B. H. Bariamov, and M. Monecke, "Determination of the charge carrier concentration and mobility in n-GaP by Raman spectroscopy", Phys. Stat. Sol. B **119**, 595 (1983). https://doi.org/10.1002/pssb.2221190219.

[1.37] B. H. Bairamov, V. V. Toporov, N. V. Agrinskaya, E. A. Samedov, G. Irmer, and J. Monecke, "Raman scattering by coupled LO-phonon-plasmon modes in CdTe", Phys. Stat. Sol. B **146**, K161 (1988). https://doi.org/10.1002/pssb.2221460254.

[1.38] Z. C. Feng, W. J. Choyke, and J. A. Powell, "Raman determination of layer stresses and strains for heterostructures and its application to the cubic 3C-SiC/Si system", J. Appl. Phys. **64**, 6827–6835 (1988). https://doi.org/10.1063/1.341997.

[1.39] W. S. Li, Z. X. Shen, Z. C. Feng, and S. J. Chua, "Temperature dependence of Raman scattering in the hexagonal gallium nitride", J. Appl. Phys. **87**, 3332–3337 (2000). https://doi.org/10.1063/1.372344.

[1.40] H.-C. Wang, Y.-T. He, H.-Y. Sun, Z.-R. Qiu, D. Xie, T. Mei, C. C. Tin, and Z. C. Feng, "Temperature dependence of Raman scattering in 4H-SiC films under different growth conditions", Chin. Phys. Lett. **32**, 047801 (2015). https://doi.org/10.1088/0256-307X/32/4/047801.

[1.41] J. B. Cui, K. Amtmann, J. Ristein, and L. Ley, "Noncontact temperature measurements of diamond by Raman scattering spectroscopy", J. Appl. Phys. **83**, 7929–7933 (1998). https://doi.org/10.1063/1.367972.

[1.42] R. Han, B. Han, M. Zhang, X. Y. Fan, and C. Li, "Temperature-dependent Raman scattering in round pit of 4H – SiC", Diam. Relat. Mater. **20**, 1282–1286 (2011). https://doi.org/10.1016/j.diamond.2011.07.009.

[1.43] B. Wang, J. Yin, D. Chen, X. Long, L. Li, H.-H. Lin, W. Hu, D. N. Talwar, R.-X. Jia, Y.-M. Zhang, I. T. Ferguson, W. Sun, Z. C. Feng, and L. Wan, "Optical and surface properties of 3C-SiC thin epitaxial films grown at different temperatures on 4H-SiC substrate", Superlattices Microstruct. **156**, 106960 (2021). https://doi.org/10.1016/j.spmi.2021.106960.

[1.44] Z. C. Feng (Ed.), Semiconductor Interfaces and Microstructures, World Scientific Publishing, Singapore, p. 328 (1992). https://doi.org/10.1142/1568.

[1.45] Z. C. Feng (Ed.), Semiconductor Interfaces, Microstructures and Devices: Properties and Application, CRC Publisher, London, p. 308 (1993).

[1.46] Z. C. Feng and R. Tsu. (Eds.), Porous Silicon, World Scientific Publishing, Singapore, p. 488 (1994).

[1.47] Z. C. Feng and J. H. Zhao (Eds.), Silicon Carbide: Materials, Processings and Devices, Taylor & Francis Books, New York, p. 416 (2003). https://doi.org/10.4324/9780203496497.

[1.48] Z. C. Feng (Ed.), SiC Power Materials – Devices and Applications, Springer, Berlin, p. 450 (2004). https://link.springer.com/book/10.1007/978-3-662-09877-6.

[1.49] Z. C. Feng (Ed.), III-Nitride Semiconductor Materials, Imperia College Press, London, p. 440 (2006). https://doi.org/10.1142/p437.

[1.50] Z. C. Feng (Ed.), III-Nitride Devices and Nanoengineering, Imperia College Press, London, p. 462 (2008). https://doi.org/10.1142/p568.

[1.51] Z. C. Feng (Ed.), Handbook of Zinc Oxides and Related Materials: Volume 1 (Materials, and Volume 2) Devices and Nano-Engineering,, CRC Press, Taylor & Francis Group, London; New York, p. 440+p. 640 (2013). https://doi.org/10.1201/b13068; https://doi.org/10.1201/b13072.

CHAPTER 2

SiC and IV-IV Semiconductors

2.1 RAMAN SPECTRA OF DIAMOND, AMORPHOUS CARBON, AND CARBON NANOTUBE

In this chapter, we present Raman scattering of carbon (C)-based materials, including diamond thin film, amorphous carbon, and carbon nanotube. Figure 2.1 exhibits a Raman spectrum of diamond thin film on Si (100) grown by the C_2H_2/O_2 flame method [2.1]. It clearly shows a sharp peak at 1332 cm^{-1}, which is characteristic of the diamond structure [2.1–2.3]. Also, it exhibits a much smaller broad peak centered at about 1556 cm^{-1}, which can be attributed to the graphitic or amorphous diamond-like carbon [2.1]. More Raman spectroscopic studies on epitaxial diamond materials can be found in [2.2, 2.3] and references therein.

The ion-implanted amorphous carbon has been studied by Raman scattering [2.4]. Raman spectrum of un-implanted amorphous carbon exhibits the well-known graphic-like D-band near 1,360 cm^{-1} and G band near 1,600 cm^{-1} [2.5, 2.6]. Efforts to synthesize the carbon-nitride phase have been explored by ion implantation into amorphous carbon.

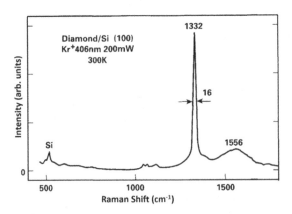

FIGURE 2.1   Raman spectrum of diamond thin film on Si (100) grown by C_2H_2/O_2 flame method.

Source: From [2.1], fig. 3, with reproduction permission of Springer.

                                                    DOI: 10.1201/9781032644912-2

With the ion dose of 5×10^{16} N+/cm², these G and D peaks became much broader. A diffuse band between 600 and 800 cm^{-1} is observed. This broad Raman peak was observed in the amorphous carbon implanted with nitride ions and was attributed to some form of disorder in the carbon film [2.4]. It is seen that at the ion doses above 5×10^{17} N+/cm² (figure 2.4(a) in [2.4]) and at a high ion implantation temperature of 800°C (figure 2.4(b) in [2.4], the G and D peaks become distinct again. This may be due to bubble or crack formation at very high ion doses, leading to less nitrogen-containing carbon compounds [2.4].

There exist different kinds of structures made from carbon, including diamond, graphite, and carbon nanotube (CNT). Since the discovery of CNT in 1991 by Iijima [2.7], CNT and related materials/structures have attracted intense research interests in their electronic, optical mechanical, and other properties that make them an interesting material for basic research and many applications. Novel properties arise from the unique one-dimensional (1D) and multi-faceted structure of CNT with unusual mechanical strength and 1D electronic transport [2.8]. These unique properties make them promise in potential use for 1D quantum wires, optical switches, nano-transistors, and other essential electronic devices. CNTs are recognized as fascinating materials to trigger a new generation of development in nano-scale devices, optical communication, carbon chemistry, and functional structural materials [2.9, 2.10]. Raman scattering has been used for intense investigation on CNTs [2.8, 2.10].

The author and collaborators have conducted research on CNT and related material [2.11, 2.12]. A typical Raman spectrum on CNT, unpublished previously, is displayed in Figure 2.2. It shows the D (1355 cm^{-1}) and G (1587 cm^{-1}) bands and their high-order combinations. For example, the peak at 2706 cm^{-1} is from the two overtones of D, the peak at 3237 cm^{-1} is 2G, the peak at 2943 cm^{-1} is D + G, and the peak at 4289 cm^{-1} is G + 2D.

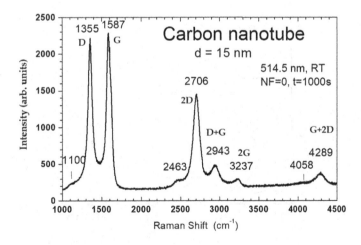

FIGURE 2.2 A typical Raman spectrum from a carbon nanotube (CNT), measured at RT and under 514.5 nm, with a time simulation of 1,000 seconds.

Source: Unpublished and original data.

2.2 RAMAN SCATTERING OF POROUS Si, C-Si, AND MICROCRYSTALLINE Si

Crystalline silicon (Si) possesses an indirect bandgap of about 1.1 eV at room temperature (RT), which can illumine in the near-infrared (NIR) region only. The discovery (1990–91) of the RT visible luminescence from porous silicon (PS) [2.13, 2.14] has raised a great deal of interest and effort on PS and related materials [2.15, 2.16]. PS can be easily prepared by electrochemical etching techniques from Si electrodes in HF:ethanol electrolytes. PS may emit strong and visible photoluminescence [2.13] or electroluminescence [2.14], which may achieve the combination of optoelectronics and integrated techniques based upon Si and open a new field in optoelectronics.

We have presented the anomalous temperature behavior of Raman spectra from visible light-emitting porous Si [2.17]. Figure 2.3 exhibits Raman spectra from typical porous Si formed by anodic oxidation in HF on p-type <111> Si, measured at 300 K (a) and (a'), and 80 K (b) and (b'). The inset part of (a') and (b') is the expansion of the major band, showing the peak of 516 cm^{-1} at 300 K and 512 cm^{-1} at 80 K, the width of 6 + 6 cm^{-1} at 300 K, and 12 + 8 cm^{-1} at 80 K, respectively. This PS Raman-active mode exhibits a downward shift in frequency and a broadening in line width when the temperature is decreased from 300 to 80 K, which is opposite to that of c-, μc-, and a-Si [2.18]. It was observed [2.18] that comparative Raman measurements on c-Si showed a peak of 517 cm^{-1} at 300 K and 520 cm^{-1} at 80 K, the width of 4 + 4 cm^{-1} at 300 K and 4 + 3 cm^{-1} at 80 K, respectively, while those on

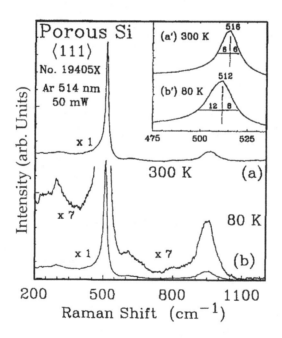

FIGURE 2.3 Raman spectra from typical porous Si formed by anodic oxidation in HF on p-type <111> Si, at 300 K (a) and (a'), and 80 K (b) and (b'). The inset part of (a') and (b') is the expansion of the major band.

Source: From [2.17], figure 1, with reproduction permission of Elsevier.

μc-Si showed a peak of 520 cm^{-1} at 300 K and 522 cm^{-1} at 80 K, the width of 20 + 10 cm^{-1} at 300 K and 12 + 8 cm^{-1} at 80 K, respectively. This anomalous T-behavior of PS Raman, in reverse of c-Si and μc-Si, can be explained by the quantum confinement and strain effect [2.17, 2.18].

Z. C. Feng et al. have performed a combined optical, surface, and nuclear microscopic assessment of PS formed in HF-acetonitrile [2.19]. Raman spectra, measured under Kr$^+$ 4067 Å and at 300 K, of porous Si prepared from different Si wafers: (a) 2 Ω-cm Si (100), (b) 0.2 Ω-cm Si (100), (c) 0.02 Ω-cm Si (111), and (d) 1Ω-cm polycrystalline Si, respectively, were presented in figure 2.4 of [2.19]. As the resistivity decreases from 2 to 0.2 Ω-cm for Si (100) and 0.02 Ω-cm for Si (111), the dominant Raman band is shifted downward, broadened, and more asymmetric. This shift or variation in Raman line shape characterizes the change in porous Si structures and is like our results on P–Si membranes prepared in ordinary HF solutions [2.17, 2.18]. The line shape analyses by way of the spatial correlation model (SCM) were given in detail [2.19].

2.3 RAMAN SCATTERING OF Ge AND SiGe

Compound $Si_{1-x}Ge_x$ belongs to group IV-IV materials and plays an important role in Si-based integrated circle technology. These materials and related structures can be easily incorporated into the well-developed Si-integrated technology, promoting the development of new-generation electronic and optoelectronic devices. The introduction of SiGe to standard Si-MOSFET technology allows bandgap engineering with enhanced performance of HMOS transistors. Si/SiGe is also one of the most promising material systems for optoelectronic devices in the important optical communication wavelength region of 1.3–1.5 μm, despite the indirect bandgap nature of Si. $Si_{1-x}Ge_x$/Si quantum structures possess many good applications for heterojunction bipolar transistors, infrared detectors using inter-subband transitions, and other devices. The growth of Si/SiGe quantum cascade (QC) structures based upon the inter-subband emission is challenging to explore the realization of Si-based QC laser. The Si/SiGe quantum well infrared photodetectors are promising for multi-spectral infrared imaging suitable for the long-, mid-, and short-wavelength infrared atmospheric windows. In addition, the Si/SiGe/Si is a promising candidate for the detection of midrange X-ray photons with better absorption coefficients [2.20 and references therein].

Figure 2.4 shows Raman spectra of sandwiched Si/SiGe/Si heterostructures, including two $Si_{1-x}Ge_x$ sandwiched between Si, measured under 632.8 nm excitation and at RT, using a Renishaw Raman microscope [2.20].

The Raman spectra of element bulk Si and Ge were previously measured under 514.5 nm excitation and at RT in figure 2.3 of [2.21]. The main Si peak is at 520 cm^{-1}, and the Si two TA phonons peaks are at 302, 227, and 434 cm^{-1}, respectively. The Ge frequency peak is at ~300 cm^{-1}. In current Figure 2.4 for $Si/Si_{1-x}Ge_x$/Si structures, the Si 520 cm^{-1} phonon mode from the Si cap and substrate is superior to all other features. The Si–Si vibration mode near 500 cm^{-1} from the $Si_{1-x}Ge_x$ layer is overwhelmed by this mode and appears as a weak shoulder of it. Raman spectra exhibit the characteristic Ge–Ge and Ge–Si vibration modes as well as the Ge–Si alloying features. The Ge–Ge and Si–Ge vibration modes from

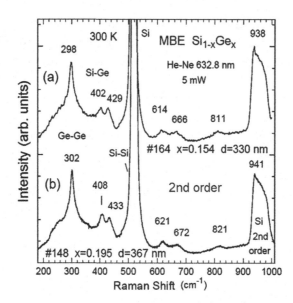

FIGURE 2.4 Raman spectra of two $Si_{1-x}Ge_x$ sandwiched between Si, measured under 632.8 nm excitation and at RT.

Source: From [2.20], figure 1, with reproduction permission of Elsevier.

$Si_{1-x}Ge_x$ are observed at near 300 and 400 cm^{-1}, with Raman shifts depending on the Ge content. The additional mode near 430 cm^{-1} and the low-energy shoulder (~250 cm^{-1}) of the Ge–Ge mode are attributed to Si–Ge ordering. Features between 920 and 1000 cm^{-1} are due to the second-order Raman features of Si. Other features located near 620, 670, and 820 cm^{-1} may come from the phonon overtones and combinations. The $Si_{1-x}Ge_x$ Raman spectral features appear clearer under the excitation of the 632.8 nm line from a He–Ne laser, as shown in Figure 2.4, than that by various green-blue lines from an Ar^+ laser, because of the thick Si caps (~150 nm). The combined investigation via visible and UV or visible and deep UV Raman scattering measurements on epitaxial $Si_{1-x}Ge_x$ layers has led to essential information on not only composition but also lattice strain [2.20].

2.4 RAMAN STUDIES OF EPITAXIAL 3C–SiC ON Si

Cubic silicon carbide (3C–SiC) is a promising material for electronic and optical devices working under high temperature, high radiation flux, and other severe environments. It possesses special properties, such as a large energy gap (2.2 eV at RT), high electric break-down field, high saturation drift velocity, moderately high electron mobility, chemical inertness, and temperature stability [2.22]. 3C–SiC has a lattice constant of 4.359 Å at RT, which is lower than that of Si, 5.430 Å at RT [2.23]. This large lattice difference of about 20% and an 8% mismatch in the thermal expansion coefficients between the two materials hindered exploring how to grow epitaxial 3C–SiC layers on Si wafers [2.23, 2.24]. However, this situation has changed due to the achievement of the successful growth of the large size of 3C–SiC films on Si wafers by way of chemical vapor deposition (CVD) by the NASA Lewis group in 1983 [2.25]. Since then, strong interest has been rekindled, and a great deal of effort and progress has been made [2.22–2.26]. Furthermore, it is due to another factor, that is, the realization and commercialization of hexagonal silicon carbide wafers, mainly 6H– and 4H–SiC wafers, and applications on SiC and related materials/devices have been greatly developed in the recent four decades.

Raman scattering technology has played an important role in the research and development (R&D) of 3C–SiC/Si [2.23, 2.24, 2.26]. The author and collaborators have conducted a variety of Raman studies on 3C–SiC/Si [2.23, 2.27–2.34]. Raman backscattering investigation for a series of cubic SiC (3C–SiC) single-crystal films grown on (100) Si by way of CVD with SiC film thicknesses d(SiC) from 0.6 to 17 μm has been performed. Raman spectra of samples with d(SiC) > 4 μm show a sharp and strong feature that obeys the selection rule for the 3C–SiC LO(Γ) phonon line. The Raman signals from the SiC film and the Si substrate show the same polarization behavior, which confirms that the crystalline orientations of the substrate and 3C–SiC film are the same [2.23, 2.27].

Figure 2.5 shows Raman spectra of four 3C–SiC/Si (100) samples with the film thickness d(SiC) from 3.2 to 11 μm, measured under 514.5 nm excitation and at RT. The intensities of 3C–SiC longitudinal optical (LO) and transverse optical (TO) phonon modes increase, with an increase of film thickness and with respect to that of the Si substrate. For the poor quality one with d(SiC) of 3.2 μm, the allowed LO mode is weaker than the forbidden TO mode. Also, some extra features exist, corresponding to other polytypes of SiC or defects. The other two 3C–SiC/Si films of (a) 4 μm and (b) 11 μm, and the 3C–SiC free film (c) 11 μm with the Si substrate removed, have the LO mode stronger than TO. Also, after the removal of the Si substrate, the LO and TO lines will shift their position a little. These Raman line shifts can be used to determine the stress and strain inside the 3C–SiC films [2.27, 2.31]. The relationship between the Raman feature and film quality can be established [2.23].

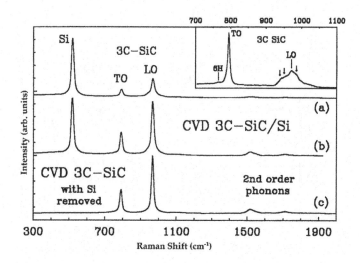

FIGURE 2.5   Raman spectra of four 3C–SiC/Si (100) with the film thickness of 3.2 μm (right-up corner), (a) 4 μm, (b) 11 μm, and (c) 11 μm, with the Si substrate removed, measured under 514.5 nm excitation and at RT.

Source: From [2.31], figure 2, with reproduction permission of Elsevier.

Raman spectra of 3C–SiC/Si (100) with a film thickness of 17 μm, in the backscattering geometry with polarization arrangements were presented in figure 2.2 of [2.23], measured under 514.5 nm excitation and at RT. For a zincblende crystal of 3C–SiC with T_d symmetry, doubly degenerate TO phonon at the Γ point is forbidden, and the singly degenerate LO phonon at the Γ point is allowed with the same polarization behavior as the incident polarization. It is identified that these selection rules are obeyed and the CVD-deposited SiC film has the same crystal orientation as the (100) Si substrate [2.23].

To study the effects of biaxial stress and strain in CVD 3C–SiC films, we etched 3C–SiC/Si samples to remove a square window in the central portion of the Si substrate, with an HF/HNO$_3$ 1:1 etch solution and a wax mask (see figure 2.1 in [2.27]). In this way, we can measure two regions of a sample, that is, the unetched 3C–SiC/Si region and the Si-etched 3C–SiC window region, by use of a micrometer screw movement, with the measurements of the two regions being made under the same conditions. Figure 2.6 shows such comparative backscattering Raman spectra of a CVD 3C–SiC/Si sample. In the perfect backscattering geometry, only the LO(Γ) phonon is allowed, and the TO(Γ) phonon is forbidden. The Raman shift difference of about 2 cm^{-1} between the two regions of the 3C–SiC film with and without the Si substrate can be clearly observed.

A set of formulas for generalized axial stress has been derived to calculate the stress and related parameters. A series of samples with the 3C–SiC film thicknesses between 4 and 17 μm was measured, the stresses and strains are calculated by us. The stresses in SiC films on Si are in the order of 10^9 dyn/cm^2 or 10^8 Pa, and the strains are 0.1–0.2% [2.27].

In many 3C–SiC/Si samples grown by CVD, some extra modes, besides the ordinary Raman phonon modes, have been observed. Figure 2.7 shows such an example, Raman

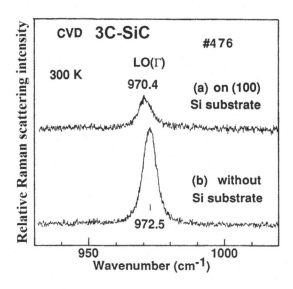

FIGURE 2.6 Comparative Raman spectra of a CVD 3C–SiC/Si sample, (a) on 3C–SiC/Si region, and (b) on 3C–SiC film region with the Si substrate removed, measured under 514.5 nm excitation and at RT, respectively.

Source: From [2.27], figure 3, with reproduction permission of AIP.

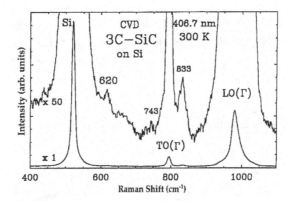

FIGURE 2.7 Raman scattering from a CVD 3C–SiC/Si(100) excited by 406.7 nm line from a Kr+ laser at 300 K. When the samples are magnified 50 times, weak modes at 620, 743 and 833 cm⁻¹ are clearly observed.

Source: **From [2.29], figure 2, with reproduction permission of Springer.**

scattering from a CVD 3C–SiC/Si(100) excited by the 406.7 nm line from a Kr+ laser at 300 K. When the samples are magnified 50 times, weak modes at 620, 743, and 833 cm⁻¹ are clearly observed.

We have investigated the optical spectral properties from three CVD 3C–SiC/Si(100) samples, with different thicknesses of 4, 7, and 16 μm [2.29]. In figure 2.3 of [2.29], besides ordinary Raman features from the Si substrate, 3C–SiC TO(Γ) and LO(Γ), additional modes at 620 and 833 cm⁻¹ are observed from the thinnest film, and for a 7 μm 3C–SiC/Si, besides the 620 cm⁻¹ mode, a mode at ~680 cm⁻¹ is seen. However, for the thickest film, features at 620, 680, and 833 cm⁻¹ are not so obvious even after magnification. Therefore, it is reasonable to correlate these modes to the defects near the 3C–SiC/Si interface region, where a high density of structural defects exists.

To understand the nature or origins of the previously observed Raman defect modes, theoretical explorations have been conducted [2.29, 2.30]. By incorporating Raman scattering data of phonons at critical points in the Brillouin zone (BZ), Talwar and Feng [2.30] have constructed a phenomenological lattice-dynamical model for perfect 3C–SiC and adopted a Green's function (GF) theory to treat the vibrations of both intrinsic and radiation-induced defects. Our calculations have also revealed a gap mode at 625 cm⁻¹ due to an isolated nitrogen defect, that is, an isolated nitrogen donor occupying the Si site in SiC, which is suggested to be responsible for the defect mode observed at 620 cm⁻¹ from our Raman measurements in Figure 2.7. It was also considered that the features at ~680 cm⁻¹ could be due to the D_I di-vacancy vibration gap mode [2.29]. The second-order Raman scattering gives an insight into the lattice-dynamical properties and ultimately information on the strength and range of interatomic forces [2.22]. Figure 2.8 shows a second-order Raman spectrum from a CVD 3C–SiC free film (16 μm).

Due to momentum conservation, the first-order Raman modes can be observed only at the center of BZ Γ point for a perfect zincblende cubic SiC. The requirement of a net zero

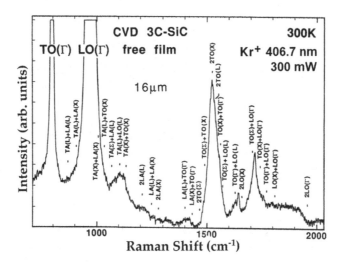

FIGURE 2.8 Second-order Raman spectrum from a CVD 3C–SiC free film, with a thickness of 16 μm, excited by 406.7 nm from a Kr⁺ laser at 300 K.

Source: From [2.22], figure 6.15, with reproduction permission of Springer.

wavevector for two phonon states leads to the observable phonon overtones and combinations from Γ, X, and L symmetry points. The zone center TO and LO phonon frequencies (at Γ) of 3C–SiC have been determined from Raman measurements [2.22, 2.23, 2.27]. The frequencies for TO and LO phonons at L were known from the phonon dispersion curves along (111) direction deduced from phonon data of different SiC polytypes. TO(X) and LO(X) values were obtained from photoluminescence (PL) measurements for 3C–SiC, which has an indirect gap and conduction minima at X [2.35].

With these data, one can obtain all possible combinations of Raman phonon modes below 2000 cm⁻¹. Our experimental Raman features at 1121, 1239, 1312, 1402, 1520, 1624, 1714, 1856, and 1896 cm⁻¹ are assigned to two-phonon Raman scattering combinations, as marked in Figure 2.8 [2.22, 2.36, and references therein]. The intensities of the second-order Raman features are more than one order of magnitude weaker than those of the first-order phonons. The strongest second-order Raman phonon mode, located at 1520 cm⁻¹, is assigned to 2TO(X). The second strongest two-phonon feature at 1714 cm⁻¹ is close to TO(X) + LO(Γ). The polarization behaviors of the first- and second-order Raman spectra from this 3C–SiC film were measured and shown in figure 6.18 in [2.22].

Windl et al. [2.37] have calculated the second-order Raman spectra of SiC, using ab initio phonon eigen solutions and phenomenological polarizability coefficients and compared them with experimental results. D. N. Talwar [2.38] performed penetrative GF theoretical calculations and obtained the triply degenerate F_2 gap modes near 630 and 660 cm⁻¹, respectively. The GF simulations of impurity vibrations for a neutral nearest-neighbor antisite SiC–C_{Si} pair-defect (C_{3v}-symmetry) provide a gap mode at ~670 cm⁻¹, between the forbidden gaps of acoustic and optical branches [2.38]. The calculated results of localized vibrational modes are compared with the Raman spectrum of Figure 2.8 (reported earlier in 2004 [2.22, 2.36]).

In addition, three representative Raman experiments and theory combination studies from our collaborative research team are briefly introduced. In the article "Adducing Crystalline Features from Raman Scattering Studies of Cubic SiC Using Different Excitation Wavelengths" [2.32], the 3C–SiC TO mode line shapes have been analyzed by the SCM and the LO modes done by the LO phonon and plasmon coupling (LOPC) according to the scattering theory, in which the doping level and the free-carrier concentration can be deduced. In the paper "Assessing Biaxial Stress and Strain in 3C–SiC/Si (001) by Raman Scattering Spectroscopy" [2.33], a series of highly strained 3C–SiC films on Si of different thicknesses (0.1–12.4 μm), prepared in a vertical chemical vapor deposition (V-CVD) by varying the growth time between 2 minutes to 4 hours, were investigated. Raman scattering data of optical phonons are carefully analyzed using an elastic deformation theory with inputs of hydrostatic-stress coefficients from a realistic lattice-dynamical approach that helped assess biaxial stress, in-plane tensile, and normal compressive strain. Conventional elastic deformation theory was used to derive the necessary expressions involving stress coefficients – correlating them with Raman phonon shifts and hydrostatic, uniaxial, and biaxial stresses. In [2.34], conventional elastic deformation theory was used for incorporating optical phonon shifts of "as-grown" and "free-standing" films to appraise the crystalline quality of material samples and estimate stress/strains in 3C–SiC films. High-density intrinsic defects present in 3C–SiC films and/or epilayer/substrate interface are likely to be responsible for releasing misfit stress/strains.

2.5 RAMAN STUDIES OF EPITAXIAL 3C–SiC ON 4H–SiC

Different from 3C–SiC/Si, with a large mismatch of about 20% in lattice constant and 8% in the thermal expansion coefficients between the two materials, mentioned in the last section, there exists a good match in lattice constant and thermal expansion coefficient for the 3C–SiC/4H–SiC heterostructure. 3C–SiC is the cubic structure with a 2.3-eV band gap, whereas 4H–SiC is hexagonal, possessing a 3.2-eV band gap (both at RT) [2.39]. The conduction band offset between them is as large as near 1 eV, and two-dimensional electron gas can be induced by the spontaneous field in C-face 4H SiC [2.39, 2.40], which provides great potential applications to novel high electron mobility devices [2.40]. Due to the lattice mismatch ratio between Si substrate (0.5431 nm at RT) and 3C–SiC film (0.436 nm at RT) of about 20%, and the difference of thermal expansion coefficient of nearly 8% (Si is 3×10^{-6} K^{-1}; SiC is 2.77×10^{-6} K^{-1}), there exist many defects and high residual stress in 3C–SiC films [2.41]. To obtain better quality 3C–SiC films, 4H–SiC substrates were employed for the growth of 3C–SiC [2.39–2.41]. A series of 3C–SiC on 4H–SiC using high-temperature chemical vapor deposition have been grown and characterized by X-ray diffraction (XRD), atomic force microscopy, scanning electron microscopy (SEM), transmission electron microscopy (TEM), X-ray photoelectron spectroscopy (XPS), PL, and Raman spectroscopy. The influences of epitaxial growth temperature on surface/structural properties were studied. The optimized growth temperature was found through our penetrative and delicate measurements and analyses. The structural properties are well elucidated for the defect formation, especially the double positioning boundary defects, in the CVD-grown 3C–SiC/4H–SiC samples [2.39–2.41].

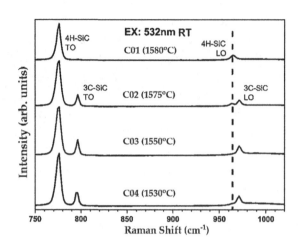

FIGURE 2.9   Raman scattering spectra at RT and under 532 nm excitation for four 3C/4H–SiC with different epitaxial growth temperatures.

Source: From [2.41], figure 4(c), with reproduction permission of Elsevier.

Here, typical Raman spectral studies on some CVD-grown 3C–SiC/4H–SiC samples are presented [2.41]. Figure 2.9 shows Raman scattering spectra, measured at RT and under 532 nm excitation, for four 3C/4H–SiC with different epitaxial growth temperatures of 1580°C (C01), 1575°C (C02), 1550°C (C03), and 1530°C (C04), respectively.

For the highest growth temperatures of 1580°C (C01), only substrate 4H–SiC TO mode at near 776 cm^{-1} and LO mode at 964 cm^{-1} are clearly observed. As the epitaxial growth temperature was decreased to 1575°C (C02), by only 5°C, the 3C–SiC TO at 796 cm^{-1} and LO at 971 cm^{-1} appeared obviously, while the 4H–SiC LO was depressed relatively. With a further decrease of the epitaxial growth temperature to 1550°C (C03) and 1530°C (C04), 3C–SiC TO and LO further increased in intensity relatively, while 4H–SiC mode is further weakened relatively. In combination with other measurements of XRD, XPS, and UV (325 nm)–DUV (266 nm) Raman, we obtained the optimized epitaxial growth temperature of 1530°C.

Figure 2.10 exhibits Raman scattering spectra at RT of the 3C/4H–SiC sample C04 (1530°C) excited by different excitations of (a) 532, (b) 325, and (c) 266 nm laser, respectively. The spectrum (a), as mentioned earlier, displays the 4H–SiC TO from the substrate strongly, and 3C–SiC film TO and LO relatively weaker. This is because the 532 nm laser line has a large penetrating depth through 3C–SiC and 4H–SiC [2.22, 2.24]. While using the UV 325 nm excitation, the penetration depth in 3C–SiC is 1.5 μm [2.22, 2.24], which is shallower than the film thickness of 4.6 μm [2.41], and, therefore, it can't detect Raman signals from the 4H–SiC substrate. Under the DUV 266 nm excitation, the laser light penetration depth in 3C–SiC is further less (about 0.2 μm) [2.41] and unable to detect Raman signals from the 4H–SiC substrate too. So, in Figure 2.10(b) and (c), only 3C–SiC TO and LO features are clearly recorded.

Figure 2.11 presents the temperature-dependent Raman spectra (TDRS) of 3C–SiC/4H–SiC C03 sample excited by 325 nm laser in the temperature range of 300 K–750 K. Three

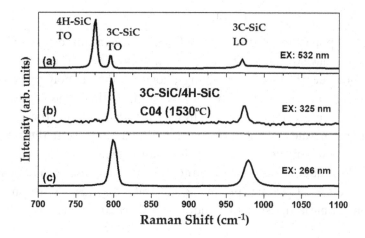

FIGURE 2.10 The Raman scattering spectra (measured at RT) of the 3C/4H–SiC sample C04 excited by different excitation of (a) 532, (b) 325, and (c) 266 nm laser, respectively.

Source: From [2.41], figure 3, with reproduction permission of Elsevier.

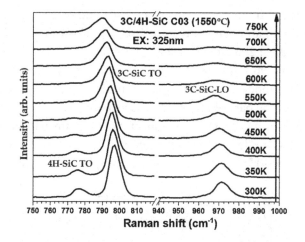

FIGURE 2.11 Temperature-dependent Raman spectra of 3C–SiC/4H–SiC C03 sample excited by 325 nm laser in the temperature range of 300–750 K.

Source: From [2.41], figure 5, with reproduction permission of Elsevier.

peaks at 300 K are 4H–SiC TO at 776 cm^{-1}, 3C–SiC TO at 798 cm^{-1} and 3C–SiC LO at 971 cm^{-1}, respectively. With the increase of temperature, the TO peak of 4H–SiC from 776 cm^{-1} gradually moves to the direction of low wave number, and its peak strength gradually weakens. It almost disappeared at 600 K. The variation of 3C–SiC TO peak positions, obtained by Gaussian fits, in the temperature range of 300 K to 750 K, can be described by Eq. (1.28) [1.41, 1.42, 2.42]. The full width at half maximum (FWHM) values can be fitted by Eq. (1.29) [1.42, 2.43]. The obtained values of 3C–SiC TO frequencies and FWHMs, as well as their variations, are given and discussed in detail [2.41].

2.6 RAMAN SCATTERING OF 4H–SiC

Among about 250 polytypes of silicon carbide (SiC), several important polytypes of SiC such as 4H and 6H have C_{6v} crystallographic symmetry. In the "a" direction, 4H–SiC and 6H–SiC are almost identical with <1% change; however, the 4H polytype consists of four units in the c-direction, and the 6H consists of six units. Different polytypes have different band gaps, electron mobility, and other physical properties. Since the late 1980s, high-quality and large-size wafers of both 6H– and 4H–SiC have come into the industry production. Gradually, the wafer size has become bigger and bigger, up to 8 inches recently, and the wafer crystalline quality has improved, and the wafers have been widely used for device applications. Wafers of SiC are also promising substrates for nitride semiconductor growth due to their compatible lattice structure and similar thermal expansion coefficients, as well as for other materials.

4H–SiC has attracted significant attention due to its wide band gap (3.2 eV at RT), high electron mobility, and excellent thermal properties [2.44]. Raman spectroscopy has been widely used for the characterization and investigation of 4H–SiC materials [2.43–2.49]. The author (Z. C. Feng) and collaborators have conducted various Raman studies on 4H–SiC [2.50–2.59], with some typical Raman spectra of 4H–SiC displayed in Figure 2.12.

Figure 2.12 shows a Raman spectrum of semi-insulating 4H–SiC taken at RT under 514.5 nm excitation, with an enlarged one in Y-scale to show detail. The 4H–SiC characteristic modes of E_2(TA) at 205.5 cm^{-1}, E_1(LA) at 268.1 cm^{-1}, A_1(LA) at 610.0 cm^{-1}, E_2(TO) at 776.5 cm^{-1}, E_1(TO) at 796.9 cm^{-1}, and A_1(LO) at 967.3 cm^{-1} modes are observed, respectively. Here the E_1(TO) mode is just a forbidden mode, while a weak Raman doublet at about 633 cm^{-1} may also be part of the overtone spectrum [2.53].

Figure 2.13 exhibits second-order Raman spectra of semi-insulating 4H–SiC and n-type doped 4H– and 6H–SiC, under the same conditions as Figure 2.12. The nominal nitrogen concentration was 2.1×10^{18} cm^{-3} for both doped samples. All the values of peaks a–g in

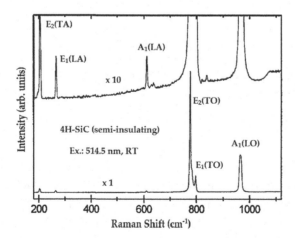

FIGURE 2.12   Raman spectrum of semi-insulating 4H–SiC taken at RT under 514.5 nm excitation, with an enlarged Y-scale to show details.

Source: From [2.53], figure 1, with reproduction permission of APS.

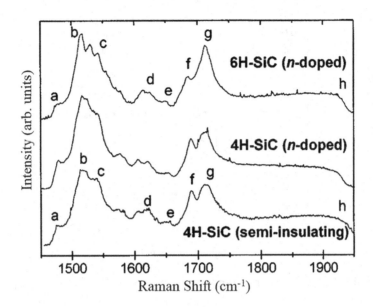

FIGURE 2.13 Second-order Raman spectra, at RT and 514.5 nm excitation of semi-insulating 4H–SiC and n-type doped 4H– and 6H–SiC. The nominal nitrogen concentration was 2.1×10^{18} cm^{-3} for both doped samples.

Source: From [2.53], figure 2, with reproduction permission of APS.

Figure 2.13 are listed in a table of [2.53]. As predicted in the last section for Raman scattering of 3C–SiC, the strongest second-order Raman phonon mode is located at 1520 cm^{-1}, assigned to 2TO(X), and the second strongest two-phonon feature at 1714 cm^{-1} is close to TO(X) + LO(Γ). Here for the second-order Raman scattering for 4H–SiC and 6H–SiC in Figure 2.13, the strongest second-order Raman peak "b" is located at 1515–1516 cm^{-1}, close to above 2TO(X) from 3C–SiC, while the second strongest two-phonon feature "g" is located at 1713–1714 cm^{-1}, close to above TO(X) + LO(Γ) from 3C–SiC too.

2.7 RAMAN SCATTERING OF 6H–SiC

A series of Raman studies on 6H–SiC were conducted by us [2.44, 2.52, 2.60–2.64]. We have employed the UV 413 nm excitation and made a combined Raman and luminescence assessment of epitaxial 6H–SiC films grown on 6H–SiC, measured at 80 K, showing the Ti-related A-, B-, and C-series PLs and outgoing resonance Raman scattering from 6H–SiC [2.60]. The effects of ion implantation on 6H–SiC by Raman scattering were studied [2.61, 2.62, 2.64]. Figure 2.14 shows micro-Raman spectra from three n–/n+ epitaxial 6H–SiC of co-implanted C–Al ions at RT with different concentrations of (a) E6, C$^+$, and Al$^+$ of 8×10^{20} cm^{-3}; (b) F3, C$^+$, and Al$^+$ of 1×10^{21} cm^{-3}; (c) F5, C$^+$, and Al$^+$ of 2×10^{21} cm^{-3}; and an annealed sample AF5 (annealed from F5 under 1500°C for 30 minutes).

The spectrum for the non-implanted corner exhibits the main Raman modes for a perfect 6H–SiC crystal [2.44, 2.52]. After C$^+$–Al$^+$ co-implantation at RT with different concentrations, three broad Raman bands appeared, with the band center at near 500, 800, and 1420 cm^{-1}, respectively, as shown in Figure 2.14(a)–(c). They are caused by Si–Si, Si–C, and

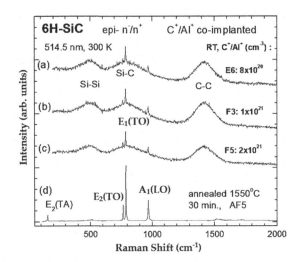

FIGURE 2.14 Raman spectra, under 514 nm excitation, from (a)–(c) three n–/n+ epitaxial 6H–SiC with C–Al ions co-implanted at RT with different concentrations and (d) an annealed sample.

Source: From [2.64], figure 3, with reproduction permission of Elsevier.

C–C vibrations, respectively, which are typical for amorphous SiC [2.64]. Single crystalline 6H–SiC characteristic Raman bands [2.44, 2.52] of two E_2(TO) and one A_1(LO) are weakly superposed on the top of the Si–C amorphous band. These indicate the damage of 6H–SiC crystallinity and the formation of the amorphous phase due to ion implantation. After the samples are annealed at 1550°C for 30 minutes, these broad amorphous-like features are almost eliminated, and a Raman spectrum like those from a single crystalline 6H–SiC appears as shown in Figure 2.14(d), with the first-order peaks of 6H–SiC E_2(TO) (767 cm^{-1}), E_1(TO) (797 cm^{-1}), and A_1(LO) (965.5 cm^{-1}) (2.44, 2.52, 2.65, 2.66), indicating the recovery of the SiC crystalline structure.

2.8 ANGLE-DEPENDENT RAMAN SCATTERING OF 4H–SiC AND ISOTROPY

Angle-dependent Raman scattering (ADRS) or rotation Raman scattering has been applied to investigate the epitaxial GaN (see Section 3.7) and crystalline 4H–SiC [2.59]. We employed the ADRS to study the phonon anisotropy property of the wurtzite 4H–SiC crystal both theoretically and experimentally. The angle-dependent polarized Raman scattering measurements were performed on the a-face 4H–SiC crystal and comparative c-face 4H–SiC, by adjusting the polarized vector of the incident and scattered laser light. Corresponding Raman selection rules are derived according to measured scattering geometries to illustrate the angle dependence. The angle-dependent intensities of phonon modes are calculated and compared to the experimental scattering intensities, yielding the Raman tensor elements of A_1, E_1, and E_2 phonon modes. These detailed theoretical calculation results on the Raman selection role, Raman tensor elements, and the variations of Raman spectral intensities of wurtzite 4H–SiC are in good coincidence with experimental

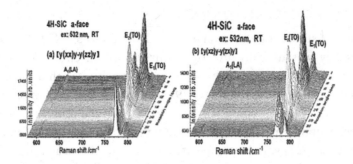

FIGURE 2.15 The rotation Raman spectra at the a-plane of 4H–SiC, with the angle varied from 0° to 360° (step size of 5°), under (a) parallel and (b) perpendicular polarization, respectively.

Source: From [2.59], figure 3, with reproduction permission of free open-assess MDPI Materials.

data. To our knowledge, ADRS has been applied to AlN [2.67, 2.68], MoS$_2$, and WS$_2$ [2.69], and Raman tensors for these wurtzite wide-gap materials are also obtained.

Figure 2.15 shows the rotation Raman spectra at the a-plane (1120) of 4H–SiC, in the wavenumber range of 580–820 cm^{-1} and with the angle rotated from 0° to 360° (step size of 5°), under (a) parallel and (b) perpendicular polarization, respectively. Within Figure 2.15, the A$_1$ (LA), E$_2$ (TO), and E$_1$ (TO) modes are observed at 610.1, 776.5, and 788.1 cm^{-1}, respectively. The intensity and phase variations of these Raman modes and their dependences on the rotation angle are clearly shown. Further, there exists a shoulder peak in the close-right of E$_2$ (TO) at ~783 cm^{-1}, which is assigned as the A$_1$ (TO) mode of 4H–SiC [2.59]. In addition, a single-point scan Raman spectrum of the a-face 4H–SiC was measured and showed E$_1$ (TA) at 266 cm^{-1}, E$_1$ (TA) at 610 cm^{-1}, E$_2$ (TO) at 775 cm^{-1}, E$_1$ (TO) at 788 cm^{-1}, A$_1$ (LO) at 976 cm^{-1}, and LOPC broadband in the right side of A$_1$(LO) [2.44, 2.53, 2.59].

The isotropy characteristics were confirmed from the c-face 4H–SiC, while the anisotropy characteristic properties were revealed from the a-face 4H–SiC. The variation functional relationship of the Raman phonon modes versus the angle between the incident laser polarization direction and the parallel (perpendicular) polarization direction was obtained. By way of the parametrization on incident light and scattered light polarization vectors, the selection rules of wurtzite SiC are calculated and well established. Corresponding Raman selection rules are derived according to measured scattering geometries to illustrate the angle dependence. Based upon the selection rules, the intensity variations of the A$_1$, E$_2$, and E$_1$ modes dependent on the rotation angle are calculated, and the Raman tensor elements of various modes are well-deduced [2.59].

In figure 2.4 of [2.59], the calculated Raman intensities of A$_1$, E$_1$, and E$_2$ phonon modes versus rotation angle (0°–360°) under parallel [y(xx)y–y(zz)y] and perpendicular [y(xz) y–y(zx)y] polarization states are displayed and compared with experimental data. These detailed theoretical calculation results on the Raman tensor elements, and the variations of Raman spectral intensities, as well as the phonon anisotropy properties of wurtzite 4H–SiC, are matched well with experimental data. The Raman tensor elements of various modes are well-deduced.

2.9 TEMPERATURE-DEPENDENT RAMAN SCATTERING OF 4H–SiC

We have performed studies of TDRS on 4H–SiC from 80 to 550 K [2.56, 2.57]. Confocal Raman Microscopy was used to investigate three 4H–SiC epilayers grown on n-type 4H–SiC substrates by low-pressure chemical vapor deposition with different silane (SiH_4) to propane(C_3H_8) gas flow ratios between 0.19 and 0.44 [2.57]. The phonon frequency $\omega(T)$ and linewidth dependence of temperature (T) were analyzed and fitted with Eqs. (1.24–1.27). The experimental Raman spectra of $A_1(LO)$ mode from 4H–SiC films and $A_1(LOPC)$ mode from the substrate at variable temperatures have been fitted as a sum of $A_1(LO)$ mode and $A_1(LOPC)$ mode, where the $A_1(LO)$ mode with a Lorentz function and the $A_1(LOPC)$ mode from the substrate fitted by Eqs. (1.19–1.23). The T-dependent frequencies of the $E_2(TO)$, $E_1(TO)$, and $A_1(LO)$ modes originate from contributions of the linear thermal expansion, three-phonon decay, and four-phonon decay. With the increasing T, Raman peaks of $E_2(TO)$, $E_1(TO)$, and $A_1(LO)$ modes shift to lower frequencies. The linewidth of the $E_2(TO)$ mode decreases with the increase of the Si:C ratio and reaction gas concentration. A shorter phonon lifetime of 4H–SiC results from epilayer quality degradation and energy transfer. The $A_1(LOPC)$ mode peaks showed abnormal temperature behavior, blueshift at low temperatures, and redshift at relatively high temperatures.

A TDRS study on bulk 4H–SiC has been conducted by us [2.56], which is in a book chapter [2.44]. Three bulk 4H–SiC samples of different doping levels were prepared: undoped, medium-doped, and heavy-doped, with carrier concentrations of 9.09×10^{14}, 2.18×10^{18}, and 4.87×10^{18} cm^{-3}, respectively. A confocal micro-Raman spectroscope system, Jobin Yvon T64000 [1.20], with the 532 nm excitation from a Nd:YOV$_4$ 532 nm laser, was employed to study the Raman mode shift of bulk 4H–SiC samples with temperatures from 80 to 873 K. TDRS spectra of each sample were displayed in figure 2.1 of [2.56] and figure 2.6 of [2.44], like Figure 2.11. Monotonous downshifts of 4H–SiC $E_2(TO)$ and $E_1(TO)$ modes with increasing T were clearly observed and well fitted using Eqs. (1.24–1.26).

For Raman $A_1(LO)$ mode at about 970 cm^{-1}, its variation of peak position with temperature for an undoped sample is also monotonous and can be well fitted by Eqs. (1.19–1.23). However, anomalous variation of the $A_1(LO)$ mode peak in two doped 4H–SiC was observed, which forms the LOPC mode due to the coupling between LO phonons and plasma. Two methods, one-mode and two-mode methods, were applied to analyze the variation of the $A_1(LO)$ mode. We have demonstrated that the two-mode method is untenable in interpreting anomalous variation of $A_1(LO)$ mode with temperature, which gives rise to contradiction to temperature properties of phonon scattering modes, while the one-mode simulation is in good accordance with experimental results. Therefore, $A_1(LO)$ mode in doped 4H–SiC is changed to LOPC mode when $A_1(LO)$ mode shows asymmetric broadening. The non-monotonous variation of blue-red shifts with temperature for LOPC mode from doped 4H–SiC could be explained by the influence of the ionization process of impurities on the process of Raman scattering [2.56].

REFERENCES

[2.1] J. Hwang, K. Zhang, B. S. Kwak, A. Erbil, and Z. C. Feng, "Growth of textured diamond films on Si (100) by C_2H_2/O_2 flame method", J. Mater. Res. **5**, 2334–2336 (1990). https://doi.org/10.1557/JMR.1990.2334.

[2.2] Q. Wei, G. Niu, R. Wang, G. Chen, F. Lin, X. Zhang, Z. Zhang, and H.-X. Wang, "Heteroepitaxy of single crystal diamond on Ir buffered KTaO3 (001) substrates", Appl. Phys. Lett. **119**, 092104 (2021). https://doi.org/10.1063/5.0045886.

[2.3] Z. Shi, Q. Yuan, Y. Wang, K. Nishimura, G. Yang, B. Zhang, N. Jiang, and H. Li, "Optical properties of bulk single-crystal diamonds at 80–1200 K by vibrational spectroscopic methods", Materials **14**, 7435 (2021). https://doi.org/10.3390/ma14237435.

[2.4] D. H. Lee, B. Park, Z. C. Feng, D. B. Poker, L. Riester, and J. E. E. Baglin, "Surface hardness enhancement in ion-implanted amorphous carbon", J. Appl. Phys. **80**, 1480–1484 (1996). https://doi.org/10.1063/1.363017.

[2.5] S. Choi, S. K. Lee, N.-H. Kim, S. Kim, and Y.-N. Lee, "Raman spectroscopy detects amorphous carbon in an enigmatic egg from the upper cretaceous wido volcanics of South Korea", Front. Earth Sci. **7**, 349 (2020). https://doi.org/10.3389/feart.2019.00349.

[2.6] C. C. Zhang, S. Hartlaub, I. Petrovic, and B. Yilmaz, "Raman spectroscopy characterization of amorphous coke generated in industrial processes", ACS Omega **7**, 2565–2570 (2022). https://doi.org/10.1021/acsomega.1c03456.

[2.7] S. Iijima, "Helical microtubules of graphitic carbon", Nature **354**, 56–58 (1991). https://doi.org/10.1038/354056a0.

[2.8] M. S. Dresselhaus, A. Jorio, M. Hofmann, G. Dresselhaus, and R. Saito, "Perspectives on carbon nanotubes and graphene Raman spectroscopy", Nano Lett. **10**, 751–758 (2010). https://doi.org/10.1021/nl904286r.

[2.9] J. Zaumseil, "Luminescent defects in single-walled carbon nanotubes for applications", Adv. Opt. Mater. **10**, 2101576 (2022). https://doi.org/10.1002/adom.202101576.

[2.10] F. L. Sebastian, N. F. Zorn, S. Settele, S. Lindenthal, F. J. Berger, C. Bendel, H. Li, B. S. Flavel, and J. Zaumseil, "Absolute quantification of sp^3 defects in semiconducting single-wall carbon nanotubes by Raman spectroscopy", J. Phys. Chem. Lett. **13**, 3542–3548 (2022). https://doi.org/10.1021/acs.jpclett.2c00758.

[2.11] Z. C. Feng, B. Xue, P. Chen, J. Lin, W. Lu, N. Li, and I. T. Ferguson, "Optical and structural studies of copper nanoparticles and microfibers produced by using carbon nanotube as templates", Proc. SPIE Nanophotonic Mater. III **6321**, 63210H-1–8 (2006). https://doi.org/10.1117/12.677666.

[2.12] Z. C. Feng, Y. Z. Huang, J. H. Ting, L. C. Chen, and W. Lu, "Field emission properties of multi-wall carbon nanotubes", Proc. SPIE **7037**, 7037OR-1–6 (2008). https://doi.org/10.1117/12.795563.

[2.13] L. T. Canham, "Silicon quantum wire array fabrication by electrochemical and chemical dissolution of wafers", Appl. Phys. Lett. **57**, 1046–1048 (1990). https://doi.org/10.1063/1.103561.

[2.14] V. Lehmann and U. Gosele, "Porous silicon formation: A quantum wire effect", Appl. Phys. Lett. **58**, 856–858 (1991). https://doi.org/10.1063/1.104512.

[2.15] Z. C. Feng and R. Tsu (Eds.), Porous Silicon, World Scientific Publishing, Singapore, p. 488 (1994).

[2.16] S. N. Agafilushkina, O. Žukovskaja, S. A. Dyakov, K. Weber, V. Sivakov, J. Popp, D. Cialla-May, and L. A. Osminkina, "Raman signal enhancement tunable by gold-covered porous silicon films with different morphology", Sensors **20**, 5634 (2020). https://doi.org/10.3390/s20195634.

[2.17] Z. C. Feng, J. R. Payne, and B. C. Covington, "Anomalous temperature behavior of Raman spectra from visible light emitting porous Si", Solid State Commun. **87**, 131–134 (1993). https://doi.org/10.1016/0038-1098(93)90341-J.

[2.18] Z. C. Feng, A. T. S. Wee, and K. L. Tan, "Surface and optical analysis of porous silicon membranes", J. Phys. D Appl. Phys. **27**, 1968–1975 (1994). https://doi.org/10.1088/0022-3727/27/9/024.

[2.19] Z. C. Feng, J. W. Yu, K. Li, Y. P. Feng, K. R. Padmanabhan, and T. R. Yang, "Combined optical, surface and nuclear microscopic assessment of porous silicon formed in HF-acetonitrile", Surf. Coat. Technol. **200**, 3254–3260 (2006). https://doi.org/10.1016/j.surfcoat.2005.07.025.

[2.20] Z. C. Feng, J. W. Yu, J. Zhao, T. R. Yang, R. P. G. Karunasiri, W. Lu, and W. E. Collins, "Optical and materials properties of sandwiched Si/SiGe/Si heterostructures", Surf. Coat. Technol. **200**, 3265–3269 (2006). https://doi.org/10.1016/j.surfcoat.2005.07.027.

[2.21] T. R. Yang, M. M. Dvoynenko, Z. C. Feng, and H. H. Cheng, "Raman spectroscopy of self-assembled Ge islands on Si", Eur. Phys. J. B **31**, 41–45 (2003). https://doi.org/10.1140/epjb/e2003-00006-x.

[2.22] Z. C. Feng, "Optical and Interdisciplinary Analysis of Cubic SiC Grown on Si by Chemical Vapor Deposition", Chapter 6 in Z. C. Feng (Ed.), SiC Power Materials – Devices and Applications, Springer, Berlin, pp. 209–276 (2004). https://doi.org/10.1007/978-3-662-09877-6_6.

[2.23] Z. C. Feng, A. J. Mascarenhas, W. J. Choyke, and J. A. Powell, "Raman scattering studies for chemical vapor deposited 3C-SiC films on (100) Si", J. Appl. Phys. **64**, 3176–3186 (1988). https://doi.org/10.1063/1.341533.

[2.24] J. A. Powell, P. Pirouz, and W. J. Choyke, "Growth and Characterization of Silicon Carbide Polytypes for Electronic Applications", Chapter 11 in Z. C. Feng (Ed.), Semiconductor Interfaces, Microstructures and Devices: Properties and Applications, Institute of Physics Publishing, Bristol, pp. 257–293 (1993).

[2.25] S. Nishino, J. A. Powell, and H. A. Will, "Production on large-scale single-crystal wafers of cubic SiC for semiconductor devices", Appl. Phys. Lett. **42**, 460–462 (1983). https://doi.org/10.1063/1.93970.

[2.26] V. Scuderi, M. Zielinski, and F. La Via, "Impact of doping on cross-sectional stress assessment of 3C-SiC/Si heteroepitaxy", Materials **16**, 3824 (2023). https://doi.org/10.3390/ma16103824.

[2.27] Z. C. Feng, W. J. Choyke, and J. A. Powell, "Raman determination of layer stresses and strains for heterostructures and its application to the cubic 3C-SiC/Si system", J. Appl. Phys. **64**, 6827–6835 (1988). https://doi.org/10.1063/1.341997.

[2.28] Z. C. Feng, C. C. Tin, R. Hu, and J. Williams, "Raman and Rutherford backscattering analyses of cubic SiC thin films grown on Si by vertical chemical vapor deposition", Thin Solid Films **266**, 1–7 (1995). https://doi.org/10.1016/0040-6090(95)06599-7.

[2.29] Z. C. Feng, D. Talwar, and I. Ferguson, "Spectroscopic Properties of Cubic SiC on Si", in S. E. Saddow, N. S. Saks, D. J. Larkin, and A. Schöner (Eds.), Silicon Carbide – Materials, Processing and Devices (MRS Symposium Proceedings Vol. **742**), Materials Research Society, K2.14.1–6 (2002). https://doi.org/10.1557/PROC-742-K2.14.

[2.30] D. N. Talwar and Z. C. Feng, "Understanding spectroscopic phonon-assisted defect features in CVD grown 3C-SiC/Si(100) by modeling and simulation", Comput. Mater. Sci. **30**, 419–424 (2004). https://doi.org/10.1016/j.commatsci.2004.02.035.

[2.31] Z. C. Feng, "Optical properties of cubic SiC grown on Si substrate by chemical vapor deposition", Microelectron. Eng. **83**, 165–169 (2006). https://doi.org/10.1016/j.mee.2005.10.044.

[2.32] S. Chen, X. Jiang, C.-C. Tin, L. Wan, and Z. C. Feng, "Adducing crystalline features from Raman scattering studies of cubic SiC using different excitation wavelengths", J. Phys. D Appl. Phys. **50**, 115102 (2017). https://doi.org/10.1088/1361-6463/aa5626.

[2.33] D. N. Talwar, L. Wan, C.-C. Tin, and Z. C. Feng, "Assessing biaxial stress and strain in 3C-SiC/Si (001) by Raman scattering spectroscopy", J. Mater. Sci. Eng. **6**, 1000324-1–8 (2017). https://doi.org/10.4172/2169-0022.1000324.

[2.34] D. N. Talwar, L. Wan, C.-C. Tin, H.-H. Lin, and Z. C. Feng, "Spectroscopic phonon and extended x-ray absorption fine structure measurements on 3C-SiC/Si (001) epifilms", Appl. Surf. Sci. **427**, 302–310 (2018). http://dx.doi.org/10.1016/j.apsusc.2017.07.266.

[2.35] W. J. Choyke, Z. C. Feng, and J. A. Powell, "Low temperature photoluminescence studies of CVD grown cubic SiC on Si", J. Appl. Phys. **64**, 3163–3175 (1988). https://doi.org/10.1063/1.341532.

[2.36] Z. C. Feng, "Second order Raman scattering of cubic silicon carbide", Sci. Access **2** (1), 242–243 (2004).

[2.37] W. Windl, K. Karch, P. Pavone, O. Schütt, D. Strauch, W. H. Weber, K. C. Hass, and L. Rimai, "Second-order Raman spectra of SiC: Experimental and theoretical results from *ab initio* phonon calculations", Phys. Rev. B **49**, 8764–8767 (1994). https://doi.org/10.1103/PhysRevB.49.8764.

[2.38] D. N. Talwar, "Probing optical, phonon, thermal and defect properties of 3C – SiC/Si (001)", Diam. Relat. Mater. **52**, 1–10 (2015). https://doi.org/10.1016/j.diamond.2014.11.011.

[2.39] B. Xin, Y.-M. Zhang, H.-M. Wu, Z. C. Feng, H.-H. Lin, and R.-X. Jia, "Kinetic mechanism of V-shaped twinning in 3C/4H-SiC heteroepitaxy", J. Vac. Sci. Tech. A **34**, 031104 (2016). https://doi.org/10.1116/1.4947601.

[2.40] H.-H. Lin, B. Xin, Z. C. Feng, and I. T. Ferguson, "Cubic SiC Grown on 4H-SiC: Growth and Structural Properties", Chapter 7 in Z. C. Feng (Ed.), Handbook of Silicon Carbide Materials and Devices, CRC/Taylor & Francis, pp. 173–196, www.routledge.com/9780367188269.

[2.41] B. Wang, J. Yin, D. Chen, X. Long, L. Li, H.-H. Lin, W. Hu, D. N. Talwar, R.-X. Jia, Y.-M. Zhang, I. T. Ferguson, W. Sun, Z. C. Feng, and L. Wan, "Optical and surface properties of 3C-SiC thin epitaxial films grown at different temperatures on 4H-SiC substrate", Superlattices Microstruct. **156**, 106960 (2021). https://doi.org/10.1016/j.spmi.2021.106960.

[2.42] J. B. Cui, K. Amtmann, J. Ristein, and L. Ley, "Noncontact temperature measurements of diamond by Raman scattering spectroscopy", J. Appl. Phys. **83**, 7929–7933 (1998). https://doi.org/10.1063/1.367972.

[2.43] R. Han, B. Han, M. Zhang, X. Y. Fan, and C. Li, "Temperature-dependent Raman scattering in round pit of 4H – SiC", Diam. Relat. Mater. **20**, 1282–1286 (2011). https://doi.org/10.1016/j.diamond.2011.07.009.

[2.44] I. T. Ferguson, Z. R. Qiu, L. Wan, J. Yiin, B. Klein, and Z. C. Feng, "Multiple Raman Scattering Spectroscopic Studies of Crystalline Hexagonal SiC Crystals", Chapter 9 in Z. C. Feng (Ed.), Handbook of Silicon Carbide Materials and Devices, CRC/Taylor & Francis, London, pp. 219–248 (2023).

[2.45] D. W. Feidman, J. H. Parker, Jr., W. J. Choyke, and L. Patric, "Phonon dispersion curves by Raman scattering in SiC, polytypes 3C, 4H, 6H, 15R, and 21R", Phys. Rev. **173**, 787–793 (1968). https://doi.org/10.1103/PhysRev.173.787.

[2.46] H. Harima, S.-I. Nakashima, and T. Uemura, "Raman scattering from anisotropic LOphonon – plasmon – coupled mode in ntype 4H– and 6H – SiC", J. Appl. Phys. **78**, 1996 (1995). https://doi.org/10.1063/1.360174.

[2.47] N. Piluso, M. Camarda, and F. La Via, "A novel micro-Raman technique to detect and characterize 4H-SiC stacking faults", J. Appl. Phys. **116**, 163506 (2014). https://doi.org/10.1063/1.4899985.

[2.48] T. Liu, Z. Xu, M. Rommel, H. Wang, Y. Song, Y. Wang, and F. Fang, "Raman characterization of carrier concentrations of Al-implanted 4H-SiC with low carrier concentration by photo-generated carrier effect", Crystals **9**, 428 (2019). https://doi.org/10.3390/cryst9080428.

[2.49] Y. Chang, A. Xiao, R. Li, M. Wang, S. He, M. Sun, L. Wang, C. Qu, and W. Qiu, "Angle-resolved intensity of polarized micro-Raman spectroscopy for 4H-SiC", Crystals **11**, 626 (2021). https://doi.org/10.3390/cryst11060626.

[2.50] C. C. Tin, R. Hu, J. Liu, Y. Vohra, and Z. C. Feng, "Raman microprobe spectroscopy of low-pressure-grown 4H-SiC epilayers", J. Cryst. Growth **158**, 509–513 (1996). https://doi.org/10.1016/0022-0248(95)00463-7.

[3.51] Z. C. Feng, A. Rohatgi, C. C. Tin, R. Hu, A. T. S. Wee, and K. P. Se, "Structural, optical and surface science studies of 4H-SiC epilayers grown by low pressure chemical vapor deposition", J. Electron. Mater. **25**, 917–923 (1996). https://doi.org/10.1007/BF02666658.

[2.52] J. C. Burton, L. Sun, M. Pophristic, F. H. Long, Z. C. Feng, and I. Ferguson, "Spatial characterization of doped SiC wafers by Raman spectroscopy", J. Appl. Phys. **84**, 6268–6273 (1998). https://doi.org/10.1063/1.368947.

[2.53] J. C. Burton, L. Sun, F. H. Long, Z. C. Feng, and I. Ferguson, "First- and second-order Raman scattering from semi-insulating 4H-SiC", Phys. Rev. B **59**, 7282–7284 (1999). https://doi.org/10.1103/PhysRevB.59.7282.

[2.54] Z. C. Feng, F. Yan, W. Y. Chang, J. H. Zhao, and J. Lin, "Optical characterization of ion implanted 4H-SiC", Mater. Sci. Forum **389–393**, 647–650 (2002). https://doi.org/10.4028/www.scientific.net/MSF.389-393.647.

[2.55] Q. Xu, H. Y. Sun, C. Chen, L.-Y. Jang, E. Rusli, S. P. Mendis, C. C. Tin, Z. R. Qiu, Z. Wu, C. W. Liu, and Z. C. Feng, "4H-SiC wafers studied by X-ray absorption and Raman scattering", Mater. Sci. Forum **717–720**, 509–512 (2012). https://doi.org/10.4028/www.scientific.net/ MSF.717-720.509.

[2.56] H. Y. Sun, S.-C. Lien, Z. R. Qiu, H. C. Wang, T. Mei, C. W. Liu, and Z. C. Feng, "Temperature dependence of Raman scattering in bulk 4H-SiC with different carrier concentration", Opt. Exp. **21**, 26475–26482 (2013). https://doi.org/10.1364/OE.21.026475.

[2.57] H.-C. Wang, Y.-T. He, H.-Y. Sun, Z.-R. Qiu, D. Xie, T. Mei, C. C. Tin, and Z. C. Feng, "Temperature dependence of Raman scattering in 4H-SiC films under different growth conditions", Chin. Phys. Lett. **32**, 047801 (2015). https://doi.org/10.1088/0256-307X/32/4/047801.

[2.58] D.-S. Zhao, F.-Z. Wang, L.-Y. Wan, Q.-Y. Yang, Z. C. Feng, "Raman scattering study on anisotropic property in wurtzite 4H-SiC", J. Light Scatter. **30**, 37–42 (2018). https://doi.org/10.13883/j. issn1004-5929.201802007.

[2.59] Z. C. Feng, D. Zhao, L. Wan, W. Lu, V. Saravade, J. Yiin, B. Klein, and I. T. Ferguson, "Angle dependent Raman scattering studies on anisotropic properties of crystalline hexagonal 4H-SiC", Materials **15**, 8751 (2022). https://doi.org/10.3390/ma15248751.

[2.60] Z. C. Feng, C. C. Tin, R. Hu, and K. T. Yue, "Combined Raman and luminescence assessment of epitaxial 6H-SiC films grown on 6H-SiC by low pressure vertical chemical vapor deposition", Semicond. Sci. Tech. **10**, 1418–1422 (1995). https://doi.org/10.1088/0268-1242/10/10/018.

[2.61] Z. C. Feng, I. Ferguson, R. A. Stall, K. Li, Y. Shi, H. Singh, K. Tone, J. H. Zhao, A. T. S. Wee, K. L. Tan, F. Adar, and B. Lenain, "Effects of Al-C ion-implantation and annealing in epitaxial 6H-SiC studied by structural and optical techniques", Mater. Sci. Forum **264–268**, 693–696 (1998). https://doi.org/10.4028/www.scientific.net/msf.264-268.693.

[2.62] Z. C. Feng, S. J. Chua, Z. X. Shen, K. Tone, and J. H. Zhao, "Microscopic probing of Raman scattering and photoluminescence on C-Al ion co-implanted 6H-SiC", Mater. Sci. Forum **338–342**, 659–662 (2000). https://doi.org/10.4028/www.scientific.net/MSF.338-342.659.

[2.63] Z. C. Feng, S. J. Chua, G. A. Evans, J. W. Steeds, K. P. J. Williams, and G. D. Pitt, "Micro-Raman and photoluminescence study on n-type 6H-SiC", Mater. Sci. Forum **353–356**, 345–348 (2001). www.scientific.net/MSF.353-356.345.

[2.64] Z. C. Feng, S. C. Lien, J. H. Zhao, X. W. Sun, and W. Lu, "Structural and optical studies on ion-implanted 6H-SiC thin films", Thin Solid Films **516**, 5217–5226 (2008). https://doi. org/10.1016/j.tsf.2007.07.094.

[2.65] Y. L. Tu, Z. C. Feng, L. Y. Jang, and W. Lu, "Synchrotron radiation X-ray absorption fine-structure and Raman scattering studies on 6H-SiC materials", SPIE **7425**, 74250Y1–9 (2009). https://doi.org/10.1117/12.825811.

[2.66] N. Daghbouja, B. S. Lib, M. Callistic, H. S. Sena, J. Lind, X. Oud, M. Karlike, and T. Polcar, "The structural evolution of light-ion implanted 6H-SiC single crystal: Comparison of the effect of helium and hydrogen", Acta Mater. **188**, 609–622 (2020). https://doi.org/10.1016/j. actamat.2020.02.046.

[2.67] W. Zheng, R. S. Zheng, H. L. Wu, and F. D. Li, "Strongly anisotropic behavior of A_1(TO) phonon mode in bulk AlN", J. Alloys Compd. **584**, 374–376 (2014). https://doi.org/10.1016/j. jallcom.2013.09.102.

[2.68] L. Jin, H. L. Wu, Y. Zhang, Z. Y. Qin, Y. Z. Shi, H. J. Cheng, R. S. Zheng, W. H. Chen, "The growth mode and Raman scattering characterization of m-AlN crystals grown by PVT method", J. Alloys Compd. **824**, 153935 (2020). https://doi.org/10.1016/j.jallcom.2020.153935.

[2.69] W. Zheng, J. Yan, F. Li, and F. Huang, "Raman spectroscopy regulation in van der Waals crystals", Photonics Res. **6**, 991–995 (2018). https://doi.org/10.1364/PRJ.6.000991.

CHAPTER 3

GaN Semiconductors

3.1 RAMAN STUDIES OF EPITAXIAL GaN ON SAPPHIRE

Since the later years of the past century, research and development (R&D) in GaN-based materials and devices have gained great achievements and opened a new echo in science and technology [1.49, 1.50, 1.52, 1.53]. Raman spectroscopy is a powerful technology in the characterization and investigation of GaN-based materials and structures. My collaborators and I have performed and published a variety of Raman scattering studies on GaN and related materials [3.1–3.19]. Raman scattering is continually acting as an important tool in the research frontiers on GaN [3.20–3.27].

We have demonstrated [3.1, 3.2] that ordinary Raman scattering in the backscattering geometry has limited use for the conventional characterization of GaN thin layers grown on sapphire substrate in industrial mass production because the strong Raman features from the sapphire substrate overwhelm the GaN Raman signals from the epitaxial film. We have described two technical methods of overcoming these difficulties. Using a near-right-angle-scattering geometry arrangement for a laser Raman system with an ordinary optical focusing lens or using a micro-Raman system with a microscope lens focusing, we can easily obtain Raman features from the thin GaN layers and greatly depress the Raman signals from the sapphire substrates. The obtained Raman spectra on GaN/sapphire exhibit all the wurtzite crystalline GaN-related Raman-active modes, including a dominant and narrow E_2 symmetric mode and $A_1(LO)$ mode, characteristic of the high quality of epitaxial GaN films grown on sapphire by the Turbo Disk MOCVD technology.

Figure 3.1 shows Raman spectra of an MOCVD-grown GaN film on sapphire, sample N423: (a) no polarization, (b) parallel-polarized, and (c) perpendicular polarized configuration (from figure 3.1 of [3.2]). Raman features from sapphire appear at 380, 418, 430, 449, 577, and 750 cm^{-1}, respectively. GaN phonons are observed for E_2 at 568 cm^{-1} and $A_1(LO)$ at 735 cm^{-1}, respectively.

Figure 3.2 exhibits Raman spectra of (a) a bare sapphire substrate and of a MOCVD-grown GaN thin (2.2 μm) film on sapphire, N667, detected with different incident angles θ of (b) 60°, (c) 85°, and (d) 89°, respectively (from figure 3.2 of [3.2]).

DOI: 10.1201/9781032644912-3

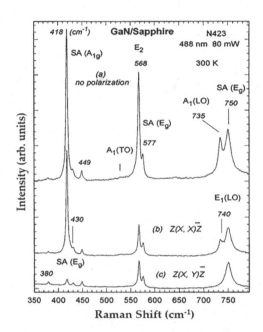

FIGURE 3.1   Raman spectra of an MOCVD-grown GaN film on a sapphire substrate, sample N423: (a) no polarization, (b) parallel-polarized, and (c) perpendicular polarized configuration.

Source: From [3.2], figure 1, with reproduction permission from AIP.

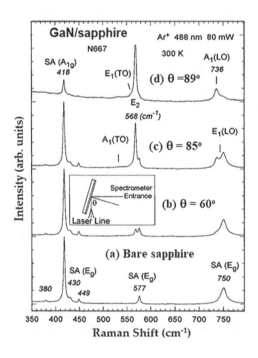

FIGURE 3.2   Raman spectra of (a) a bare sapphire substrate, and of an MOCVD-grown GaN sapphire, sample N667, with different incident angles θ of (b) 60°, (c) 85°, and (d) 89°, respectively.

Source: From [3.2], figure 2, with reproduction permission from AIP.

The Raman spectrum of bare sapphire in Figure 3.2(a) confirms the assignments in Figure 3.1. When the incident angle θ is less than 30°–40°, we obtained a spectrum like that from a bare sapphire substrate, as shown in (a). When θ is near 60°, a mode, that is, the GaN E_2, is observed weakly at 568 cm^{-1} in (b). At θ = 85°, this mode is stronger than the sapphire Eg mode at 575 cm^{-1}, and another GaN A_1(LO) mode is observed at 735 cm^{-1}, comparable in intensity to the sapphire Eg mode at 750 cm^{-1} in (c). For the case of θ = 60°, this GaN A_1(LO) mode appeared only as a low-frequency shoulder of the sapphire Eg mode at 750 cm^{-1} in (b). At θ = 89° the two GaN modes of E_2 and A_1(LO) dominated the sapphire modes in (d). A GaN E_1(TO) phonon mode is observed as a low-frequency shoulder of the strong GaN E_2 mode. The relative intensity of the sapphire strongest A_{1g} mode with respect to the GaN high E_2 mode decreases with increasing θ.

We have presented "Raman Scattering Properties of GaN Materials and Structures under Visible and Ultraviolet Excitations" [3.7]. Figure 3.3 shows Raman scattering of an undoped GaN film, 4 μm thick, grown on sapphire by MOCVD, with excitation of 514.5 nm in the backscattering geometry at room temperature (RT). Spectrum (a) measured with the incident laser light along the normal direction of the film, that is, parallel with the

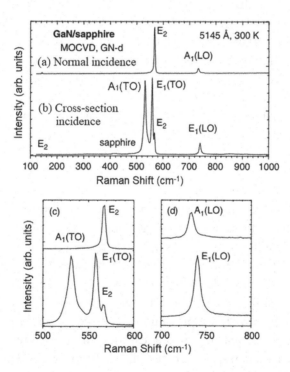

FIGURE 3.3 Raman scattering of an undoped GaN film grown on sapphire by MOCVD, at RT and with 514.5 nm excitation in the backscattering geometry. Spectrum (a) measured with the incident laser light along the normal direction of the film, that is, parallel with the c-axis of wurtzite GaN, and (b) with the incident laser light focused on the cross-section of the film, that is, perpendicular to the c-axis of GaN. Spectra (c) and (d) are the expansion regions of 500–600 and 700–800 cm^{-1}, respectively.

Source: From [3.7], figure 1, with reproduction permission from Wiley.

c-axis of wurtzite GaN, and (b) with the incident laser light focused on the cross-section of the film, that is, perpendicular to the c-axis of GaN. Spectra (c) and (d) are the expansion regions of 500–600 and 700–800 cm^{-1}, respectively (from figure 3.1 of [3.7]). A confused micro-Raman system, Renishaw M2000, with a stigmatic single spectrograph (25 cm focal length), a holographic grating (1800 or 2400 grooves per mm), notch filters, and a Peltier-cooled charge-coupled device (CCD) detector (576 by 384 pixels), was used for Raman measurements. The microscope lens focuses the laser beam on the sample surface to a spot size of 1 μm with an output power of less than 1 mW. This microscope lens also serves to efficiently collect the scattered light in the backscattering geometry. Beam splitters were employed to divide the laser and scattering beams. The experiment spectral resolutions in use were better than 1 cm^{-1} for the visible Raman measurements and 3 to 4 cm^{-1} for the UV applications (described in Section 3.5) [3.7, 3.9].

All the major first-order Raman modes, as shown in Figure 3.3, obey the selection rules of wurtzite GaN. A low E_2 symmetry mode is located at 142 cm^{-1}, a high E_2 is at 568 cm^{-1}, and the A_1 symmetry longitudinal optical (LO) phonon mode A_1(LO) is near 735 cm^{-1}. These three modes are allowed in the backscattering configuration when the laser light is incident along the normal direction of the GaN film. The transverse optical (TO) modes of A_1(TO) at 532 cm^{-1} and E_1(TO) at 557 cm^{-1} are allowed when the laser light incidence is perpendicular to the c-axis of the GaN, that is, with the backscattering measurements made on the cross-section of the film. The appearance of the E_1(LO) mode at 742 cm^{-1} is due to the so-called quasi-right-angle incidence. The observation of these Raman modes can be used to identify the wurtzite GaN. Their width and intensity can be used to assess the quality of grown materials. Their frequency position can be employed to monitor the stress and strain present in the epitaxial film [3.7, 3.9].

3.2 RAMAN SCATTERING OF Si-DOPED GaN ON SAPPHIRE

Raman scattering spectroscopic technique was employed to study a series of n-type GaN epilayers grown on c-face sapphires in an article "Material Properties of GaN Grown by MOCVD" [3.3]. The GaN films used in this section were grown on (1000)-plane sapphire substrates by MOCVD. Silicon was used as the dopant, and a typical thickness of the n-type GaN epilayers is ~2.5 μm. The doping concentrations are between 3×10^{17} and 3×10^{19} cm^{-3}, and an undoped GaN sample with a thickness of 1.5 μm and with a background free-carrier concentration below 10^{17} cm^{-3} was used for comparison. These Si-doped GaN films grown on sapphire by MOCVD were previously measured by the Hall effect and studied by way of a combined experimental and theoretical study of infrared reflectance (IRR) [3.28].

Figure 3.4 shows RT Raman spectra under 442 nm from a He–Cd laser of these samples with the doping levels marked in a figure of [3.3]. The GaN E_2 phonon mode at 568 cm^{-1} and A_1(LO) at 734 cm^{-1} are observed in the figure. The frequency of the E_2 mode in all the samples did not vary with the carrier concentration, within experimental error. The line widths of E_2 modes of the n-type GaN epilayers are almost the same, indicating similar crystal quality of the n-type GaN epilayers, although they are a little broader than that of the undoped GaN epilayer. This may be due to the increase of the doping-related defects in the n-type GaN epilayers.

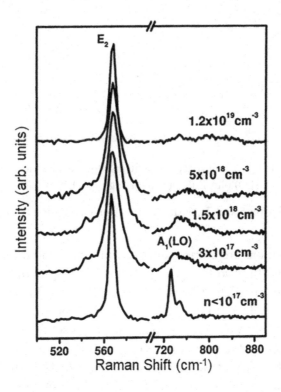

FIGURE 3.4   Raman scattering spectra at RT and 442 nm excitation of n-type GaN epilayers in the range of 500–900 cm⁻¹.

Source: From [3.3], figure 4, with reproduction permission from Wiley.

In Raman spectra, the LO phonons shifted to a higher frequency and broadened as the doping concentration increased. This phenomenon is due to the coupling between the LO phonons and damped plasmons in n-type GaN. All $A_1(LO)$ modes of these samples were theoretically fitted by Eqs. (1.19–1.23), with the fitted carrier concentration values of all samples like those obtained from Hall measurements and IRR fits [3.28], which are marked in Figure 3.4 [3.3].

Devki N. Talwar, Hao-Hsiung Lin, and Zhe Chuan Feng presented an investigation, "Anisotropic Optical Phonons in MOCVD-Grown Si-Doped GaN/Sapphire Epilayers" [3.18]. Figure 3.5 shows typical Raman scattering spectra and fittings for # S_5 sample of (a) E_2(high) mode and fitting using Eqs. (1.16–1.18) and (b) $A_1(LO)$ plus E_g(Sapphire) and fitting using Eqs. (1.19–1.23), respectively.

A comprehensive theoretical study was conducted to examine the experimental Raman vibrational properties of high-quality, Si-doped GaN/sapphire epi-films prepared by the MOCVD growth technique. In the perfect wurtzite GaN crystal, there are two hexagonally close-packed sublattices of Ga and N atomic species shifted against each other along the c-axis. An ideal structure with four atoms per unit cell belongs to the space group C_{6v} (P63mc) (see Section 2.1). At the center (i.e., $q \simeq 0$ or Γ-point) of the Brillouin zone, the group theory predicts eight sets of phonon modes given by $2A_1 + 2B_1 + 2E_1 + 2E_2$. One set

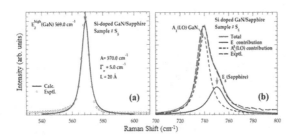

FIGURE 3.5 The Raman scattering spectra and fittings for # S_5 sample of (a) E_2(high) mode using Eqs. (1.16–1.18) and (b) A_1(LO) plus E_g(Sapphire) using Eqs. (1.19–1.23), respectively.

Source: **From [3.18], figures 5 and 7a, with reproduction permission from Elsevier.**

of A_1, E_1 optical vibrations are infrared and Raman active, and the second set is linked to the acoustic phonons. The two E_2-modes of E_2(low) and E_2(high) are also Raman active, whereas the B_1 modes are silent, that is, neither infrared nor Raman active. The E_2(high) mode in GaN corresponds to atomic oscillations in the c-plane, having a nonpolar characteristic. Consequently, the E_2(high) phonon frequency is sensitive to the residual in-plane stress. In n-GaN/sapphire the stress can be evaluated from the observed shift of E_2(high) mode with respect to a free-standing GaN film. One must also note that the A_1 and E_1 vibrations split up into LO of A_1(LO) and E_1(LO) and TO of E_1(TO) and E_1(TO) components by macroscopic electric field (i.e., LO–TO splitting). This split affects the LO phonons to have higher energies than their TO counterparts. In the heteroepitaxial growth of GaN, we have used a sapphire substrate, which exhibits seven Raman-active phonon modes expressed by $2A_{1g} + 5E_g$ [3.18].

A modified spatial correlation model (SCM) was used for simulating the Raman line shapes of the E_2(high) phonon for monitoring both the crystalline quality and the strain state. While the optical phonons in lightly doped samples are coupled to electron plasma, at higher carrier concentrations, the over-damped A_1(LO) mode vanished in the background. For each sample, we assessed the transport parameters by simulating Raman profiles of A_1(LO) line shape with contributions from plasmon–LO phonon and Lorentzian-shaped Eg sapphire mode. A realistic Green's function theory is adopted to study the vibrational modes of Si donors and Mg acceptors in GaN by including force constant changes estimated from lattice relaxations using a first-principle bond-orbital model. Theoretical results of impurity-activated modes compared favorably well with the existing Raman scattering data [3.18].

3.3 RAMAN SCATTERING ON GaN FILM CROSS-SECTION AND NONPOLAR A- AND M-PLANE GaN

We have performed Raman measurements on the cross-section of a GaN thin film, that is, a-face GaN [3.7, 3.9]. Figure 3.6 presents Raman spectra of an undoped MOCVD-grown GaN/sapphire with excitation at 514.5 nm in the backscattering geometry at RT, with incidence along the cross-section of the 4 µm-thick GaN film, i.e., perpendicular to the c-axis of GaN, (a)–(h) with the laser focusing point moved from substrate side to film surface side

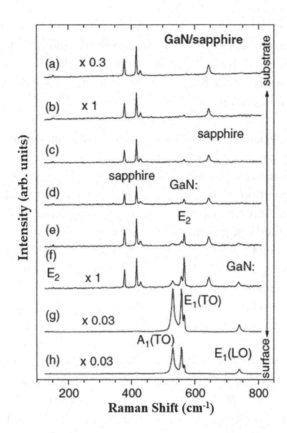

FIGURE 3.6 Raman spectra of an undoped MOCVD-grown GaN/sapphire with excitation at 514.5 nm in the backscattering geometry at room temperature, with incidence on the cross-section of the film, that is, perpendicular to the c-axis of GaN; (a)–(h) moved from substrate side to film surface side.

Source: From [3.7], figure 4, with reproduction permission from Wiley.

with 0.5 μm per detecting point. As seen in Figure 3.6(a), only Raman features from sapphire are present, indicating that the focusing spot is located at the substrate. When moved toward the GaN/sapphire interface, Raman scattering features from GaN film become stronger. When the incident laser focusing spot is away from the GaN/sapphire interface and close to the GaN film surface, only GaN features like Figure 3.3(b) are observed. The Raman spectral intensities in Figure 3.6(g) are stronger than those in Figure 3.6(h) because the latter has the focusing spot (~1 μm in diameter) partially spread outside the film surface. As demonstrated in Figure 3.3, the Raman $A_1(LO)$ mode is obvious in the case of normal incidence, whereas the $E_1(LO)$ mode is strong for cross-section incidence.

Our further careful measurements and analyses on the $E_1(LO)$ mode have revealed that with the move of the laser focusing spot from the substrate or GaN/sapphire interface side to the surface, the peak position of the $E_1(LO)$ mode is slightly shifted higher. This shift, with a total amount of 0.3 cm^{-1} from the substrate side to the surface region, may be caused by the variation of strain and/or doping level in the film growth direction [3.7, 3.9].

Nonpolar (1–120) oriented a-plane GaN is an attractive material for applications in electric and optoelectronic devices due to the elimination of the internal electric fields-induced quantum-confined Stark effect. Unfortunately, the crystalline quality of the nonpolar a-plane GaN epilayer grown on the semipolar (1–102)-oriented r-plane sapphire substrate is still much poorer than the conventional polar (0001)-oriented c-plane GaN epilayer grown on c-plane sapphire substrate. Because of the large mismatch in both lattice constant and thermal expansion coefficient between a-plane GaN epilayer and r-plane sapphire substrate, high density of basal plane stacking faults and strong crystallographic anisotropy are usually generated in the nonpolar a-plane GaN epilayer during the epitaxial growth process. In the work [3.17], the reduction in the crystalline quality anisotropy and the in-plane strain and influence on the crystalline quality for the nonpolar a-plane GaN epilayer with nano-scale island-like SiN_x interlayer grown on r-plane sapphire substrate were studied.

Figure 3.7(a) presents RT Raman spectra for two typical samples of S_0 and S_4 prepared via epitaxial lateral overgrowth (ELOG). Sample S_0 is the nonpolar a-plane GaN epilayer grown on the r-plane sapphire substrate without the nano-scale island-like SiN_x interlayer. Another sample S_4 is the nonpolar a-plane GaN epilayer grown with the nano-scale island-like SiN_x interlayer on the r-plane sapphire substrate. All the nonpolar a-plane GaN epilayer samples used in this study were grown in a low-pressure (40 Torr) metal-organic chemical vapor deposition (MOCVD) system. Trimethyl-aluminum (TMAl), trimethyl-gallium (TMGa), silane (SiH_4), and ammonia (NH_3) were used as the precursors for Al, Ga, Si, and N, respectively. Before the growth, the r-plane sapphire substrate was heated up to 1060°C in H_2 ambience to remove surface contamination. A 60-nm-thick composite AlN buffer layer, from bottom to top, consisted of a 12-nm-thick high-temperature (HT)-grown AlN nucleation layer grown at 1050°C, an 18-nm-thick low-temperature (LT) AlN nucleation layer grown at 720°C, and a 30-nm-thick HT-AlN buffer layer grown at 1050°C, was deposited on the r-plane sapphire substrate. A 1.2 µm-thick nonpolar a-plane GaN epilayer was then deposited at 1030°C with a growth rate of 1.8 µm/h over the composite AlN buffer layer. These growth and processing details are given in [3.17].

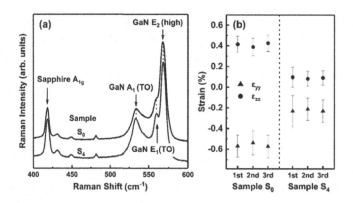

FIGURE 3.7 The Raman spectra (a) and the in-plane strains calculated with error bars along c- and m-directions (b) for samples S_0 and S_4, respectively.

Source: From [3.17], figure 3, with reproduction permission from Elsevier.

In Figure 3.7(a), the Raman peak located at 418 cm^{-1} is assigned to be the A_{1g} mode peak for the sapphire substrate. Three peaks at 532, 559, and 568 cm^{-1} are identified to be the GaN $A_1(TO)$, $E_1(TO)$, and $E_2(high)$ mode peaks, respectively.

To quantitatively characterize the evolution of the in-plane strains after the insertion of the nano-scale island-like SiN$_x$ interlayer, calculations are performed. According to the liner elasticity theory, the stress along the growth direction vanished and the strains along three axes x, y, and z in nonpolar a-plane GaN epilayer can be described as [3.17]:

$$\varepsilon_{yy} = -\frac{c_{11}}{c_{12}}\varepsilon_{xx} - \frac{c_{13}}{c_{12}}\varepsilon_{zz} \qquad (3.1)$$

Here, c_{ij} (i, j = 1, 2, 3) are the elastic stiffness constants. x, y, and z axes are defined as the nonpolar a-plane GaN (11–20) oriented a-direction, (1–100) oriented m-direction, and [0001] oriented c-direction, respectively. On the other hand, the relationship between the strains and the Raman shift is given by the following Eq. [3.17]:

$$\Delta\omega_\lambda = a_\lambda\left(1 - \frac{c_{11}}{c_{12}}\right)\varepsilon_{yy} + \left(b_\lambda - \frac{c_{13}}{c_{12}}\right)\varepsilon_{zz} + c_\lambda\left|\varepsilon_{xx} - \varepsilon_{yy}\right| \qquad (3.2)$$

where $\Delta\omega_\lambda$ is the Raman shift for certain mode, a_λ, b_λ, and c_λ are the phonon deformation potentials of the related mode. By fitting the Raman peaks with Lorentz line shape, the peak positions for the GaN $E_1(TO)$ and $E_2(high)$ modes were extracted from the Raman spectra shown in Figure 3.17(a). The strains could be obtained by solving Eqs. (3.1) and (3.2) with the Raman shift values measured for the GaN $E_1(TO)$ and $E_2(high)$ modes. The calculated in-plane strains for the nonpolar a-plane GaN epilayer samples S_0 and S_4 are displayed in Figure 3.7(b) including error bars. It was clearly demonstrated that sample S_0 suffered large compressive strain along the m-direction and tensile strain along the c-direction. However, the in-plane strains in sample S_4 were reduced significantly in both m- and c-directions due to the SiN$_x$ interlayer. The Raman spectra revealed that the SiN$_x$ interlayer plays a crucial role in compensating the in-plane strains and suppressing the crystalline quality anisotropy by triggering the ELOG of the nonpolar a-plane GaN epilayer. In fact, it was found that the in-plane compressive strain along the m-direction and the tensile strain along the c-direction could be decreased by approximately 62% and 78%, respectively, due to the insertion of the SiN$_x$ interlayer. As a result, the lattice constants along the two in-plane directions became very close to those of the strain-free GaN [3.17].

3.4 RAMAN STUDIES OF GaN ON Si

GaN thin films have also been grown on various foreign substrates, such as sapphire, GaAs, and SiC. Silicon is an attractive substrate because of its high crystal quality, large area size, low manufacturing cost, and potential application in integrated devices. Therefore, GaN and related materials and structures grown on Si are promising for developing a new generation of devices by the combination of Si- and III-N-based materials and technologies in the 21st century. However, due to the even larger difference in lattice constant and thermal expansion coefficient between GaN and the silicon substrate compared with that between GaN and sapphire, it is even more difficult to grow high-quality GaN films and structures on Si substrates than on sapphire. We have made great efforts to investigate MOCVD-grown GaN/Si (001) [3.5, 3.8]. GaN films and InGaN–GaN multiple quantum well structures have been grown on Si (001) substrate with specially designed composite intermediate layers (CIL) consisting of an ultrathin amorphous silicon layer and a GaN/$Al_xGa_{1-x}N$ multilayered buffer by low-pressure metal-organic MOCVD [3.5, 3.8]. The wurtzite crystal nature of our MOCVD-grown GaN materials on Si can be verified also via Raman scattering.

Figure 3.8(a) shows a Raman spectrum from a MOCVD-grown GaN/Si (001). A strong band is shown at 520 cm^{-1} from the Si substrate, and a band at ~300 cm^{-1} is due to the acoustic phonons of Si. There are three Raman bands representative of the wurtzite GaN phonon modes of E_2(low) at 141 cm^{-1}, E_2(high) at 567 cm^{-1}, and A_1(LO) at 736 cm^{-1} (Table 1.4). Figure 3.8(b) exhibits a fine Raman scan from another sample, between 500 and 580 cm^{-1}, and two additional wurtzite GaN modes of A_1(TO) at 532 cm^{-1} and E_1(TO) at 557 cm^{-1} (Table 1.4) are revealed. There are no phonon modes related to the cubic phase GaN to be observed in the Raman spectra of our GaN/Si samples [3.5, 3.8]. These provide confirmation together with XRD measurements on the wurtzite structural nature of our MOCVD-grown GaN materials although they were grown on cubic structural Si substrates.

We have reported "X-ray Absorption and Raman Study of GaN Films Grown on Different Substrates by Different Techniques" [3.15] and applied the SCM of Eqs. (1.16–1.18) to analyze comparatively the GaN high E_2 modes for two MOCVD-grown GaN/sapphire and two MBE-grown GaN/Si. The SCM fitting data showed that GaN films grown on sapphire possess a correlation length much longer than those grown on Si, respectively. It is confirmed by Raman scattering spectra that the phonon modes show a significant shift between different GaN epitaxial layers with different growth conditions and that the Raman technique can provide complementary information to reveal the local variation of stress in epilayers. These could help to optimize the growth parameters and obtain the relationship between the GaN film crystalline quality and the doping level.

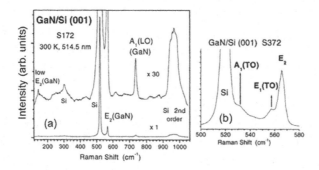

FIGURE 3.8   Raman spectra of two undoped GaN/Si (001) with CILs, with different frequency ranges.

Source: From [3.8], figure 2, with reproduction permission from Elsevier.

3.5 RESONANT RAMAN STUDIES OF GaN ON Si AND SAPPHIRE

For p-type doped GaN and GaN-based LED samples, resonance Raman scattering (RRS) with multiple LO phonon modes up to 5LO can be observed under UV 325 nm excitation. Figure 3.9 shows such an example. UV 325 nm excited RT PL exhibits the 3.4 eV GaN edge PL band, p-type doping-related 3.25 eV band, and LED emissions in 2.6 to 3.2 eV [3.8].

In addition, several sharp lines are observed between 3.3 and 3.8 eV. These are resonance Raman scattering LO modes, which were rescanned in Raman frequency in the inset of Figure 3.9, including a magnification of 10 for the low-frequency range. These RRS mLO modes are enhanced due to the resonance with the GaN fundamental 3.4-eV PL band from the 325 nm (3.6 eV) incoming laser excitation, which have been reported for GaN grown on sapphire substrates ([3.7, 3.8] and references therein). They are now observed for GaN materials and structures grown on Si substrates, with combined PL and Raman scattering under UV excitation conditions.

Figure 3.10 shows the UV 325 nm excited Raman spectra of an MOCVD-grown p-GaN/sapphire, with mLOs up to m = 6. The inset is its PL spectrum between 2.3 and 3.75 eV, showing Mg-related 2.8–3.1 eV PL transitions [3.7].

These results clearly show that high-order LOs appear due to the resonance of the Raman-scattered photons with the GaN 3.4 eV bandgap. In non-resonance excitation conditions, Raman scattering of A_1(LO) and E_2(high) phonons occurs via the deformation-potential mechanism with A_1(LO) much weaker than the E_2(high) mode. For the resonance case, A_1(LO) phonons are scattered via the Fröhlich mechanism with a strong enhancement of multiple A_1(LO) phonon scattering. When the energy of the Raman-scattered photon, E_{out}, matches the GaN bandgap, E_g, an outgoing resonance occurs with an incoming photon of energy, E_{in}, scattered m times by LO phonons:

$$E_{out} = E_{in} - m\hbar\omega_{LO} \sim Eg \qquad (3.1)$$

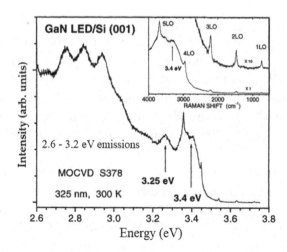

FIGURE 3.9 Combined UV excited PL and resonance Raman spectra from a GaN LED grown on Si (001) with CILs (S378).

Source: From [3.8], figure 9, with reproduction permission from Elsevier.

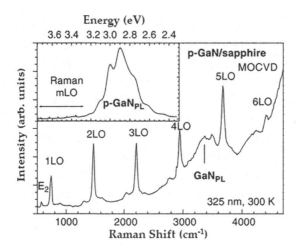

FIGURE 3.10 UV 325 nm excited Raman spectra of an MOCVD-grown p-GaN/sapphire. mLOs up to m = 6 are observed. The inset is its PL spectrum between 2.3 and 3.75 eV, showing Mg-related 2.8–3.1 eV PL transitions.

Source: **From [3.7], figure 6, with reproduction permission from Wiley.**

where $\hbar\omega_{LO}$ is the energy of an $A_1(LO)$ phonon and $\hbar$ is Planck's constant divided by 2π. From the inset in Figure 3.10, it can also be seen that strong emissions spreading between 2.5 and 3.2 eV, which are Mg impurity-related features in GaN, and weak sharp lines between 3.3 and 3.75 eV are detected at the same scan. The latter part is enlarged in Figure 3.10 and with the x-axis converted to Raman shift in wavenumber, as described earlier. The GaN band edge PL emission band near 3400 cm⁻¹, that is, 3.4 eV, is also detected together with sharp mLO (m = 1–6) phonon modes. This sensitive UV Raman–PL approach therefore offers a good way to characterize the p-GaN epitaxial materials [3.7, 3.9].

3.6 TEMPERATURE-DEPENDENT RAMAN SCATTERING OF GaN

Measurement of the temperature-dependent Raman line shifts and widths allows comparison with predictions of anharmonic crystal theory. The knowledge of the Raman shift with temperature would allow its use as an in-situ temperature diagnostic tool [3.4, 3.6]. Figure 3.11 shows temperature-dependent Raman spectra (TDRS) of a GaN thin film, undoped and 2 μm thick, grown on sapphire by low-pressure metal-organic chemical vapor deposition (LP-MOCVD). The orientation of the c-axis was normal to the surface of the GaN film [3.4].

Raman spectra were recorded using a SPEX 1704 spectrometer equipped with a microscope attachment and a liquid-nitrogen-cooled CCD under the 488 nm excitation of an Ar-ion laser. The Raman spectral resolution is better than 0.5 cm⁻¹. and the power of the laser on the sample was set to about 2 mW. The sample was kept in a Linkam hot/cold stage, which was placed under the microscope of the Raman system, and used for the measurement. Because both the GaN film and sapphire substrate are transparent, the temperature deviation due to laser heating should be negligible. The temperature was measured by a K-type thermocouple with an accuracy of better than ±1 K [3.4, 3.6].

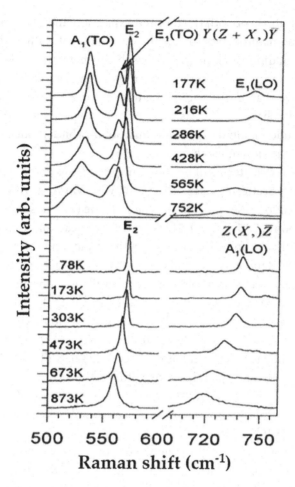

FIGURE 3.11   First-order polarized Raman spectra recorded in the temperature range T = 177–752 K and 78–873 K under two backscattering geometries, respectively. The upper spectra are in Y (Z + X) $\bar{Y}$ geometry, with X||[100], Y||[010], and Z||[001]. The lower spectra are for Z (X,) $\bar{Z}$ geometry.

Source: From [3.4], figure 1, with reproduction permission from Wiley.

The first-order polarized Raman spectra in Figure 3.11 were recorded in T = 177–752 K and 78– 873 K under two backscattering geometries, respectively. The upper spectra are in Y (Z + X) $\bar{Y}$ geometry, with X||[100], Y||[010], and Z||[001]. The lower spectra are for Z (X) $\bar{Z}$ geometry. The temperature dependence of Raman frequency ω(T) of GaN E_2 mode in Figure 3.11 can be modeled by Eqs. (1.24–1.26) for the three- and four-phonon processes [1.39, 3.11], which is displayed in figure 2 of [3.4] or figure 3 of [3.6]. The Raman linewidth can be modeled by Eq. (1.27) and exhibited in figure 3 of [3.4] or figure 4 of [3.6].

From our works [3.4, 3.6] and analyses previously, the temperature dependence of the GaN Raman modes is obtained. Both the Raman shift and linewidth exhibit a quadratic dependence on temperature in our measured temperature range. A model involving three- and four-phonon coupling was employed for the theoretical calculation of the temperature dependence of GaN Raman shift and linewidth. Our results indicate that it is necessary to include the contributions of both the thermal expansion and four-phonon terms in the four-phonon anharmonic processes to explain the change of Raman shift and linewidth with temperature. In addition, a decrease in the splitting between the LO and TO phonons with increasing temperature was observed. From these data, a weak nonlinear decrease of the transverse effective charge with increasing temperature is derived. The comparison of the transverse effective charge at RT was made between experimental data and theoretical calculations by a pseudopotential expression and bond-orbital model. A good agreement between theory and experiment is achieved [3.4, 3.6].

We conducted Raman scattering of nonpolar m-plane GaN in the article, "Temperature Dependence of Raman Scattering in m-Plane GaN with Varying III/V Ratios" [3.16]. The group theory and selection rules predict that the m-plane GaN possesses eight sets of phonon modes at the Γ point: 2A1 + 2B1 + 2E1 + 2E2. In the case of m-plane GaN and backscattering geometry, the A1 and E1 modes are both Raman and infrared active, while the two E2 modes are only Raman active, and the two B1 modes are both silent. The temperature-dependent Raman spectra of m-plane GaN on m-plane sapphire were measured from −180 to 240°C. The Raman peaks of A1(TO), E1(TO), E2(high), and E1(LO) modes shift to low frequency or redshift, which should be due to anharmonic coupling to phonons of other branches, or the heating of crystal, which cause volume expansion or lattice dilation. The temperature dependences of the Raman shift of A1(TO), E1(TO), E2(high), and E1(LO) modes from the m-plane GaN epitaxial films with different III/V ratios are well-fitted by Eqs. (1.24–1.26), combining three-phonon processes and four-phonon processes. The theoretical fitting results reveal the three-phonon processes to be dominant for the redshift of E1(LO) and E2(high) [3.16].

Our research team has recently performed "A Comparative Investigation for Optical Properties of GaN Thin Films Grown on c- and m-Face Sapphire by Metal-Organic Chemical Vapor Deposition" [3.19]. The TDRS of two GaN films grown on c-sapphire (4c) and m-sapphire (4m), with a 532 nm laser in temperatures from 80 to 800 K, are studied. Raman spectroscopy-related oscillations show systematic variations in temperature in both GaN configurations (polar and semi-polar). Following Eq. (1.28), $\omega(T) = \omega_0 - A/\{\exp[B(\hbar\omega_0/kT)]-1\}$, and Eq. (1.30), $\Delta E = 1/\tau$, the temperature dependence of the GaN E_2(high) mode is well described in detail [3.19]. The GaN thin film samples show a Raman shift toward the lower wave number and a gradual increase in FWHM with an increase in temperature. As the temperature is increased from 80 to 800 K, the biaxial structural stress in GaN switches from compressive to tensile. The stress-switch temperature point of the GaN/c-sapphire is higher than GaN/m-sapphire. The residual stress of GaN/m-sapphire is more affected by temperature, and the phonon lifetime is less affected by temperature compared to GaN/c-sapphire.

There are two other major factors that affect the phonon lifetime. One factor is the anharmonic effect of phonons, which causes one phonon to decay into several phonons. The other factor is phonon scattering caused by impurities and defects in the crystal. The phonon lifetime in HT is lower than in the LT environment. This is likely because, at HT, the interaction between optical and acoustic phonons increases, resulting in an increase in phonon scattering and a shortening of the phonon lifetime. Phonon diffusion is mainly caused by the scattering of impurity phonons and anharmonic decay. With the gradual increase in line width, the phonon lifetime decreases because of temperature-dependent variation in the probability of phonon decay to low-energy phonons. This leads to an enhancement of the phonon anharmonic decay process. For the same temperature range, the phonon lifetime of GaN/m-sapphire(4m) varies less. The phonon lifetime of (11–22) GaN films grown on m-plane sapphire is less affected by temperature compared to (0001) GaN [3.19].

3.7 ANGLE-DEPENDENT RAMAN SCATTERING OF GaN FILM

In this section, we report an investigation of the GaN phonon anisotropy employing angular-dependent Raman spectroscopy, which provides a good Raman application of significance. The hexagonal wurtzite GaN is a tetrahedrally coordinated semiconductor compound, with the space group C^4_{6v} (P6$_3$mc) of two formula units in the primitive cell. All the atoms occupy sites of symmetry C_{3v}. The following modes are Raman active: A_1(LO) = 735, A_1(TO) = 533, E_1 (LO) = 743, E_1 (TO) = 561, E_2(low) = 144, and E_2(high) = 569 cm^{-1}, respectively [3.13]. Owing to the different vibrating types, the phonon intensity induced by the electric vector will vary from the polarization of the incident laser beam and cause the manifold scattered light. The polarizability tensor of the wurtzite crystal [2.59, 2.69] has been used to verify the Raman selection rule in the experimental figure.

Experimentally, the c-axis oriented wurtzite GaN thin film used in this study is about 2 μm thick and is grown on a c-plane sapphire substrate by low-pressure metal-organic chemical vapor deposition (LP-MOCVD). Raman measurements were performed using a SPEX 1704 Raman spectrometer, equipped with a CCD detector and an Olympus microscope with a rotation stage. The 488 nm (2.54 eV) line of an Ar$^+$ laser was used as the excitation source. The polarized Raman scattering spectra were measured along the cross-section of the c-axis oriented GaN thin film as a function of the angle between the polarization direction of the incident laser with three different polarization configurations, the perpendicular, parallel, and non-polarized experiment configuration [3.13].

Under the cross-section scattering conditions, only the A_1(TO), E_1(TO), and E_2 modes are observable. The configuration of this experiment is shown in Figure 3.12(a). The ξηζ is the laboratory system while the **xyz** are congruous with the crystal axes of the single-crystal film. The **x**-axis is along the c-axis of wurtzite GaN, that is, the (0001) hexagonal direction. Figure 3.12(b) plots some typical spectra at several rotation angles.

In our experimental arrangement of Figure 3.12, the ζ axis always coincides with the **y**-axis. The rotation angle θ is between ξ and **z**-axis. Vectors $\mathbf{k}_i$ and $\mathbf{k}_s$, which are anti-parallel with each other, are the wave vector of the incident and scattered photons, respectively. The electric vector $\mathbf{e}_i$ along the ξ axis represents the polarization of the incident light. When the polarization of the scattered light ($\mathbf{e}_s$) is perpendicular to $\mathbf{e}_i$, the measured signal will be notated as ⊥, the perpendicular geometry. The parallel (//) and non-polarized (//+⊥) geometry will be noted too. Raman spectra under three polarization configurations (//+⊥,//, ⊥) were measured as functions of θ, with a data step interval of 5°. In the cross-polarized geometry, the A_1(TO) mode does not appear at θ = 0° and θ = 90° and exhibits strongly at θ = 45°, as shown in Figure 3.12(b). Raman intensities of the E_2 mode increase from θ = 0°

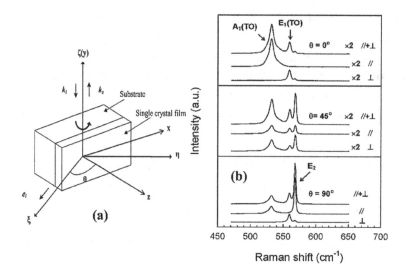

FIGURE 3.12 (a) The two coordinate systems in the experimental configuration. The $\xi\eta\zeta$ is the laboratory coordinate system, while the **xyz** coincides with the crystal axes of the sample. The **x**-axis is along the c-axis of wurtzite GaN, that is, (0001) hexagonal direction. (b) Raman spectra of the cross-section of the GaN thin film at several typical rotation angles.//, ⊥ and//+⊥ denote the parallel-, cross-, and non-polarized geometries, respectively.

Source: **From [3.13], figure 1, with reproduction permission from AIP.**

to $\theta = 90°$ in both the non- and parallel-polarized geometries. The intensities of the $A_1(TO)$ mode in the non- and parallel-polarized geometries are much stronger than that of the E_2 mode at $\theta = 0°$, while the reverse is true at $\theta = 90°$. It is clearly seen that the Raman intensities of the three modes strongly depend on the rotation angles.

Typically, according to the Raman tensors and considering the vibrating phase shift, the Raman intensity could be deduced by three cases [3.13]:

Perpendicular polarized case

$$I_{A1}{}^{\perp} \propto (\frac{|a|^2 + |b|^2}{4} - \frac{|a||b|}{2}\cos(\phi_{a-b}))\sin^2 2\theta,$$

$$I_{E1}{}^{\perp} \propto |e|^2 \cos^4\theta + |d|^2 \sin^4\theta - \frac{|e||d|}{2}\sin^2 2\theta \cos\phi_{e-d}, \quad I_{E2}{}^{\perp} \propto \frac{1}{4}|f|^2 \sin^2 2\theta \tag{3.3}$$

Parallel-polarized case

$$I_{A1}{}^{//} \propto |a|^2 \sin^4\theta + |b|^2 \cos^4\theta + \frac{|a||b|}{2}\sin^2 2\theta \cos(\phi_{a-b}),$$

$$I_{E1}{}^{//} \propto \left(\frac{|e|^2 + |d|^2}{4} + \frac{|e||d|}{2}\cos(\phi_{e-d})\right)\sin^2 2\theta, \quad I_{E2}{}^{//} \propto |f|^2 \sin^4\theta \tag{3.4}$$

Non-polarized case

$$I_{A1} \propto |a|^2 \sin^2\theta + |b|^2 \cos^2\theta, \quad I_{E1} \propto |d|^2 \sin^2\theta + |e|^2 \cos^2\theta, \quad I_{E2} \propto |f|^2 \sin^2\theta \quad (3.5)$$

where ϕ_{a-b} and ϕ_{d-e} stand for complex phases of the two independent components of the Raman tensor. Figure 3.13 shows Raman peak intensities of (a) A_1 (TO), (b) E_1 (TO), and (c) E_2 modes as a function of rotation angle (0–180°) with a data step interval of 5°, in three polarization configurations, respectively.

The theoretical fits were performed via Eqs. (3.3–3.5) and expressed by dot and solid lines in Figure 3.13, which indicate the susceptibility contribution and the phase differential of the different vibrating elements. As shown in Figure 3.13(a), the Raman intensity variation of the A_1(TO) mode with the rotating angle exhibits a sinusoidal tendency. This indicates the anisotropic feature of Raman scattering from the wurtzite GaN. It is the same for the E_1(TO) mode, as shown in Figure 3.13(b). The intensity of E_2 in Figure 3.13(c) is quite different from the A_1 and E_1 modes, because the E_2 mode has only one kind of vibrating element, while the others have two kinds and will have the phase differential contributing to the susceptibility of the polarized angle.

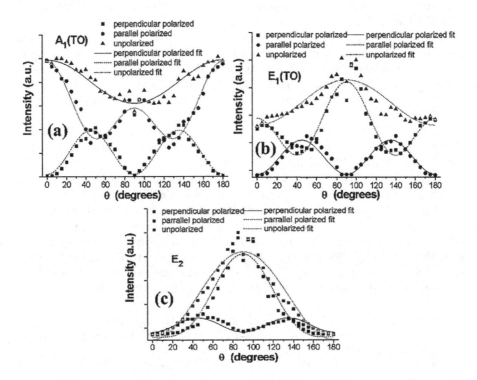

FIGURE 3.13   Raman peak intensities of (a) A_1 (TO), (b) E_1 (TO), and (c) E_2 modes as a function of rotation angle (0–180°) in three polarization configurations, respectively.

Source: From [3.13], figures 2, 3, and 4, respectively, with reproduction permission from AIP.

REFERENCES

[3.1] Z. C. Feng, M. Schurman, R. A. Stall, M. Pavloski, and A. Whitley, "Raman scattering as a characterization tool for epitaxial GaN thin films grown on sapphire by turbo disk metalorganic chemical vapor deposition", Appl. Optics **36**, 2917–2922 (1997). https://doi.org/10.1364/ao.36.002917.

[3.2] Z. C. Feng, M. Schurman, and R. A. Stall, "How to distinguish the Raman modes of epitaxial GaN with phonon features from sapphire substrate", J. Vac. Sci. Technol. A **15**, 2428–2430 (1997). https://doi.org/10.1116/1.580759.

[3.3] W. Liu, Z. C. Feng, M. F. Li, S. J. Chua, N. Akutsu, and K. Matsumoto, "Material properties of GaN grown by MOCVD", Surf. Interface Anal. **28**, 150–154 (1999). https://doi.org/10.1002/(SICI)1096-9918(199908)28:1 < 150::AID-SIA595 > 3.0.CO;2-0.

[3.4] W. S. Li, Z. X. Shen, Z. C. Feng, and S. J. Chua, "Raman scattering and transverse effective charge of MOCVD-grown GaN films between 78 and 870 K", Surf. Interface Anal. **28**, 173–176 (1999). https://doi.org/10.1002/(SICI)1096-9918(199908)28:1 < 173::AID-SIA601 > 3.0.CO;2-7.

[3.5] X. Zhang, S. J. Chua, Z. C. Feng, J. Chen, and J. Lin, "MOCVD growth and characterization of GaN films with composite intermediate layer buffer on Si substrate", Phys. Stat. Sol. (A) **176**, 605–610 (1999). https://doi.org/10.1002/(SICI)1521-396X(199911)176:1 < 605::AID-PSSA605 > 3.0.CO;2-Q.

[3.6] W. S. Li, Z. X. Shen, Z. C. Feng, and S. J. Chua, "Temperature dependence of Raman scattering in the hexagonal gallium nitride", J. Appl. Phys. **87**, 3332–3337 (2000). https://doi.org/10.1063/1.372344.

[3.7] Z. C. Feng, W. Wang, S. J. Chua, P. Zhang, K. P. J. Williams, and G. D. Pitt, "Raman scattering properties of GaN materials and structures under visible and ultraviolet excitations (invited paper)", J. Raman Spectrosc. **32**, 840–846 (2001). https://doi.org/10.1002/jrs.765.

[3.8] Z. C. Feng, X. Zhang, S. J. Chua, T. R. Yang, J. C. Deng, and G. Xu, "Optical and structural properties of GaN materials and structures grown on Si by metalorganic chemical vapor deposition", Thin Solid Films **409**, 15–22 (2002). https://doi.org/10.1016/S0040-6090(02)00096-2.

[3.9] Z. C. Feng, "Micro-Raman scattering and micro-photoluminescence of GaN thin films grown on sapphire by metalorganic chemical vapor deposition", Opt. Eng. **41**, 2022–2031 (2002). https://doi.org/10.1117/1.1489051.

[3.10] Z. C. Feng, Y. J. Sun, L. S. Tan, S. J. Chua, J. W. Yu, J. H. Chen, C. C. Yang, W. Lu, and W. E. Collins, "P-type doping in GaN through Be implantation", Phys. Stat. Sol. C **2**, 2415–2419 (2005). https://doi.org/10.1002/pssc.200461458.

[3.11] W. Tong, M. Harris, B. K. Wagner, J. W. Yu, H. C. Lin, and Z. C. Feng, "Pulse source injection molecular beam epitaxy and characterization of nano-scale thin GaN layers on Si substrates", Surf. Coat. Technol. **200**, 3230–3234 (2006). https://doi.org/10.1016/j.surfcoat.2005.07.020.

[3.12] J. W. Yu, H. C. Lin, Z. C. Feng, L. S. Wang, and S. J. Chua, "Control and improvement of crystalline cracking from GaN thin films grown on Si by metal-organic chemical vapor deposition", Thin Solid Films **498**, 108–112 (2006). https://doi.org/10.1016/j.tsf.2005.07.081

[3.13] H. C. Lin, Z. C. Feng, M. S. Chen, Z. X. Shen, I. T. Ferguson, and W. Lu, "Raman scattering study on anisotropic property of wurtzite GaN", J. Appl. Phys. **105**, 036102 (2009). https://doi.org/10.1063/1.3072705.

[3.14] Z. C. Feng, C. Tran, I. T. Ferguson, and J. H. Zhao, "Material properties of GaN films grown on SiC/SOI substrate", Mater. Sci. Forum **600–601**, 1313–1316 (2009). https://doi.org/10.4028/www.scientific.net/MSF.600-603.1313.

[3.15] Y. L. Wu, Z. C. Feng, J. F. Lee, W. Tong, B. K. Wagner, I. Ferguson, and W. Lu, "X-ray absorption and Raman study of GaN films grown on different substrates by different techniques", Thin Solid Films **518**, 7475–7479 (2010). https://doi.org/10.1016/j.tsf.2010.05.027.

[3.16] C. Chen, X. P. Shu, H. Y. Sun, Z. R. Qiu, T.-W. Liang, L.-W. Tu, and Z. C. Feng, "Temperature dependence of Raman scattering in m-plane GaN with varying III/V ratios", Adv. Mater. Res. **602–604**, 1453–1456 (2013). https://doi.org/10.4028/www.scientific.net/AMR.602-604.1453.

[3.17] J. Zhao, X. Zhang, Z. Wu, Q. Dai, N. Wang, J. He, S. Chen, Z. C. Feng, and Y. Cui, "Reduction in anisotropy and strain for non-polar a-plane GaN epi-layers with nano-scale island-like SiN_x interlayer", J. Alloys Compd. **729**, 992–996 (2017). https://doi.org/10.1016/j.jallcom.2017.09.230.

[3.18] D. N. Talwar, H.-H. Lin, and Z. C. Feng, "Anisotropic optical phonons in MOCVD grown Si-doped GaN/Sapphire epilayers", Mater. Sci. Eng. B **260**, 114615-1–10 (2020.06). https://doi.org/10.1016/j.mseb.2020.114615.

[3.19] H. Lu, L. Wang, Y. Liu, S. Zhang, Y. Yang, V. Saravade, Z. C. Feng, B. Klein, I. T. Ferguson, L. Wan, and W. Sun, "A comparative investigation for optical properties of GaN thin films grown on c- and m-face sapphire by metalorganic chemical vapor deposition", Semicond. Sci. Tech. **37**, 065021 (2022). https://doi.org/10.1088/1361-6641/ac696f.

[3.20] Y. Zeng, J. Ning, J. Zhang, Y. Jia, C. Yan, B. Wang, and D. Wang, "Raman analysis of E_2(High) and A_1(LO) phonon to the stress-free GaN grown on sputtered AlN/graphene buffer layer", Appl. Sci. **10**, 8814 (2020). https://doi.org/10.3390/app10248814.

[3.21] Z. Zhang, Z. Xu, Y. Song, T. Liu, B. Dong, J. Liu, and H. Wang, "Interfacial stress characterization of GaN epitaxial layer with sapphire substrate by confocal Raman spectroscopy", Nanotechnol. Precis. Eng. **4**, 023002 (2021). https://doi.org/10.1063/10.0003818.

[3.22] A. J. E. N'Dohi, C. Sonneville, L. V. Phung, T. H. Ngo, P. De Mierry, E. Frayssinet, H. Maher, J. Tasselli, K. Isoird, F. Morancho, Y. Cordier, and D. Planson, "Micro-Raman characterization of homoepitaxial n doped GaN layers for vertical device applications", AIP Adv. **12**, 025126 (2022). https://doi.org/10.1063/5.0082860.

[3.23] Y. Peng, W. Wei, M. F. Saleem, K. Xiao, Y. Yang, Y. Yang, Y. Wang, and W. Sun, "Resonant Raman scattering in boron-implanted GaN", Micromachines **13**, 240 (2022). https://doi.org/10.3390/mi13020240.

[3.24] P. Loretz, T. Tschirky, F. Isa, J. Patscheider, M. Trottmann, A. Wichser, J. Pedrini, E. Bonera, F. Pezzoli, and D. Jaeger, "Conductive n-type gallium nitride thin films prepared by sputter deposition", J. Vac. Sci. Technol. A **40**, 042703 (2022). https://doi.org/10.1116/6.0001623.

[3.25] S. Rao, E. D. Mallemace, G. Faggio, M. Iodice, G. Messina, and F. G. D. Corte, "Experimental characterization of the thermo-optic coefficient vs. temperature for 4H-SiC and GaN semiconductors at the wavelength of 632 nm", Sci. Rep. **13**, 10205 (2023). https://doi.org/10.1038/s41598-023-37199-6.

[3.26] B. Han, M. Sun, Y. Chang, S. He, Y. Zhao, C. Qu, and W. Qiu, "Raman characterization of the in-plane stress tensor of gallium nitride", Materials **16**, 2255 (2023). https://doi.org/10.3390/ma16062255.

[3.27] M. Elhajhasan, W. Seemann, K. Dudde, D. Vaske, G. Callsen, I. Rousseau, T. F. K. Weatherley, J.-F. Carlin, R. Butté, N. Grandjean, N. H. Protik, and G. Romano, "Optical and thermal characterization of a group-III nitride semiconductor membrane by microphotoluminescence spectroscopy and Raman thermometry", Phys. Rev. B **108**, 235313 (2023). https://doi.org/10.1103/PhysRevB.108.235313.

[3.28] Y. T. Hou, Z. C. Feng, S. J. Chua, M. F. Li, N. Akutsu, and K. Matsumoto, "Influence of Si-doping on the characteristics of GaN on sapphire by infrared reflectance", Appl. Phys. Lett. **75**, 3117–3119 (1999). https://doi.org/10.1063/1.125249.

CHAPTER 4

Other III-Nitride Semiconductors

4.1 RAMAN STUDIES OF EPITAXIAL InN ON SAPPHIRE

We have performed Raman studies for indium nitride (InN) thin films and related materials [4.1–4.4]. InN has the lowest energy band gap among III-nitrides, about 0.65 eV at RT, the lowest effective mass for electrons, high mobility, and high saturation velocity [4.4, 4.5]. Raman scattering has also played an important role in the research and exploration of InN in frontiers [4.6–4.9].

Heteroepitaxial growth of m-plane (10$\underline{1}$0) InN film on (100) γ-LiAlO$_2$ (LAO) substrate has been realized by plasma-assisted molecular beam epitaxy (MBE). Polarized Raman spectra on the m-plane, c-plane, and mixture are shown in Figure 4.1, taken in normal backscattering. Two pronounced modes located at 449 cm^{-1}, A$_1$(TO) mode in sample D (m-plane InN) and 492 cm^{-1} E$_2$(high) mode in sample C (c-plane InN), are obtained in the spectra, but both A$_1$(TO) and E$_2$(high) modes are observed in sample B (mixture). A broadened mode centered at ~584 cm^{-1} in all films is related to A$_1$(LO) or E$_1$(LO) modes [4.1]. This result is consistent with the Raman selection rule. Therein, E$_2$(high) mode can completely disappear, but A$_1$(TO) mode appears with maxima intensity under x(z,z) x-scattering configuration, where z ∥ c for nonpolar wurtzite films. In contrast, only E$_2$(high) mode can be observed for the polar film under z(x,x)z-scattering configuration. In addition, compared to free-standing InN bulk, both A$_1$(TO) and E$_2$(high) modes show blue shift with ~2 cm^{-1}, which indicates slightly residual compressive stress of ~0.2 GPa in the films [4.1].

Figure 4.1(b) shows a quarter pole figure plotted with integrated intensities of E$_2$(high) mode normalized with each intensity maxima. The intensity of E$_2$(high) mode varies with angle and is proportional to $\sin^4 \theta$ as a constant for nonpolar and polar directions in wurtzite crystals, respectively. Herein, the angle θ is set as the angle between laser polarization and c-axis direction of the wurtzite crystals. The Raman intensity trends of both m- and c-plane InN films follow the calculated Raman intensities well [4.1].

 DOI: 10.1201/9781032644912-4

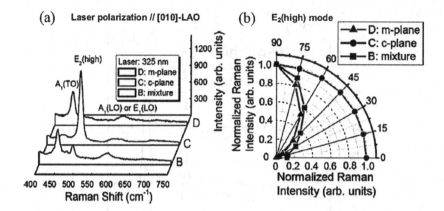

FIGURE 4.1 (a) Polarized UV Raman spectra of InN films measured at 0° with backscattering configuration; that is, the laser polarization is parallel to (010) direction of the LAO substrate. The Raman spectra are matched to x(z,z)x and z(x,x)z scattering configurations for m- and c-plane InN, respectively. (b) Integrated Raman intensities of E_2(high) mode with angle dependence plotted as a quarter pole figure. The m-plane InN shows strong angle dependence.

Source: **From [4.1], figure 5, with reproduction permission from AIP.**

In an article "Raman Scattering And Rutherford Backscattering Studies On InN Films Grown By Plasma-Assisted Molecular Beam Epitaxy", we have reported a Raman spectral investigation of InN films grown by MBE, together with the Rutherford backscattering (RBS) [4.2].

Figures 4.2–4.4 show Raman spectra of InN films grown with plasma power at 200, 300, 350, and 400 W, respectively. Typical InN Raman modes of E_2 (low) at ~88 cm⁻¹, E_2 (high) at~492 cm⁻¹, and A_1(LO) at~590 cm⁻¹ were observed from their RT Raman spectra under the 532 nm excitation [4.2]. As seen in Figure 4.2, the intensity of A_1(LO) mode with respect to E_2(high) is increased gradually with an increase in the plasma power. From Figure 4.3, it is seen that the InN Raman mode frequencies of E_2 (low) and E_2 (high) are only slightly varied with the change of plasma power.

Besides these four strong active Raman modes of E_2(low), E_2(high), and A_1(LO), in Figures 4.2 and 4.3, other two active Raman modes, A_1(TO) and E_1(TO), can be found from our measured Raman spectra. An A_1(TO) band at ~446 cm⁻¹ is observed in Figure 4.2 for five Raman spectra. The E_1(TO) is located at ~476 cm⁻¹ as a shoulder of the low-frequency side of the strong E_2(high) peak in Figure 4.3(b). It is seen that it is relatively weaker for the lowest plasma power, that is, with the lowest carrier concentration, and it became stronger for higher plasma power or higher carrier concentration. Therefore, we recognize this mode at ~476 cm⁻¹ in all Raman spectra of Figure 4.3(b) as a mixed mode of E_1(TO) and L⁻.

But the A_1(LO) mode peak frequency varies in a special way with the plasma power as shown in Figure 4.4: it increases from 200 to 230 W and further to 300 W, and then decreases

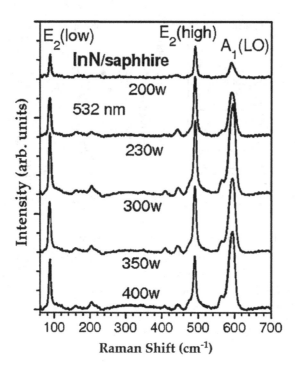

FIGURE 4.2   Raman spectra of InN films grown with plasma power at 200, 300, 350, and 400 W, respectively.

Source: From [4.2], figure 3, with reproduction permission from Elsevier.

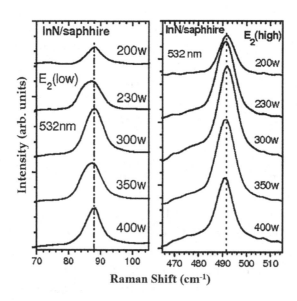

FIGURE 4.3   Raman spectra for E_2 (low) and E_2 (high) modes of InN films grown with plasma power at 200, 230, 300, 350, and 400 W, respectively.

Source: From [4.2], figure 4, with reproduction permission from Elsevier.

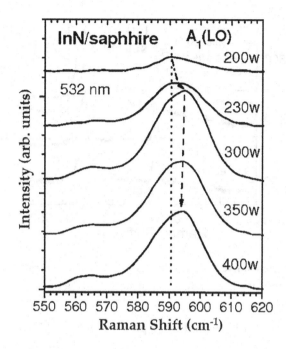

FIGURE 4.4   Raman spectra for A_1 (LO) mode of InN films grown with plasma power at 200, 230, 300, 350, and 400 W, respectively.

Source: From [4.2], figure 5, with reproduction permission from Elsevier.

to 350 W and to 400 W. The A_1(LO) mode peak exhibits a large shift among ~30 cm^{-1}, far beyond the measurement error bar. By careful watching of these five Raman spectra in Figure 4.4, we suggest that the band spreading over 575–610 cm^{-1} consists, indeed, of two modes: A_1(LO) located at about 590–591 cm^{-1} and E_1(LO) at about 594–595 cm^{-1}.

With the variation of the plasma power, that is, the carrier concentration, these two modes are competing. At the lowest plasma power (200 W) case, that is, with the lowest carrier concentration, the A_1(LO) mode is dominated. As with the slightly higher plasma power (230 W) case, that is, with the second-lowest carrier concentration, the E_1(LO) mode increases in intensity slightly to make the high-frequency wing increase in intensity. For three high plasma power (300, 350, and 400 W) cases, that is, with high carrier concentration beyond 5×10^{19} cm^{-3}, the E_1(LO) mode becomes dominant.

Also, the LO phonon–plasmon coupling (LOPC) may contribute to this region of Raman bands, which is supported by the experimental fact of Figure 4.2. At the lowest plasma power, that is, with the lowest carrier concentration, the E_2(high) mode is stronger than the A_1(LO) [plus E_1(LO)]. As increasing the carrier concentration (through increasing the plasma power), the intensity of A_1(LO) + E_1(LO) mixed mode is quickly increasing with respect to the E_2(high) mode and is superior to the E_2(high) mode for the three high plasma power (300, 350 and 400 W) cases, that is, with high carrier concentration beyond 5×10^{19} cm^{-3}. It is very reasonable to assume that the enhanced LO phonon–plasmon coupling via increasing the carrier concentration through increasing the plasma power level

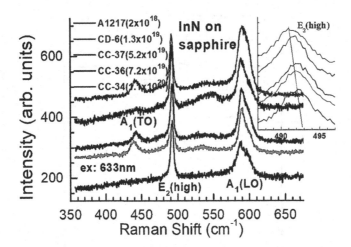

FIGURE 4.5 Raman spectra of five InN epi-films with carrier concentrations between 2×10^{18} and 1×10^{20} cm^{-3}, under the 633-nm excitation, with the inset showing the small shifts of E$_2$(high) mode with carrier concentration.

Source: From [4.4], figure 11, with reproduction permission from AIP.

makes the A$_1$(LO) + E$_1$(LO) mixed band quickly increase in intensity with respect to the E$_2$(high) band [4.2].

From our recent research "Optical, Surface and Structural Studies of InN Thin Films Grown on Sapphire by Molecular Beam Epitaxy" [4.4], Figure 4.5 shows Raman spectra of five InN epi-films with carrier concentrations between 2×10^{18} and 1×10^{20} cm^{-3}, under the 633-nm excitation, with the inset showing the small shifts of E$_2$(high) mode with carrier concentration.

It has been reported [4.4, 4.6] that the E$_2$(high) phonon mode is sensitive to biaxial strain, as a good probe for biaxial strain in the layers and correlated with the carrier density in InN materials. The inset of Figure 4.5 presents the expanded Raman spectra of E$_2$ (high) mode with different carrier concentrations, that is, doping level, under 633 nm excitation. It shows that E$_2$(high) clearly broadens and shifts to lower frequencies as the doping level increases. This correlation between the E$_2$(high) frequency and carrier density is well corresponding to the Hall measurements. The correlation observed between the E$_2$(high) frequency and carrier density, therefore, indicates that donor-type nitrogen vacancies associated with threading dislocations also play a role in determining the background electron density in InN layers.

4.2 RAMAN INVESTIGATION OF EPITAXIAL AlN ON SAPPHIRE AND SiC

We have performed a series of Raman scattering studies on aluminum nitride (AlN) nanowire (NW) [4.10], AlN epitaxial films [4.11–4.16], and bulk [4.17]. AlN is an ultrawide-bandgap semiconductor with the largest direct band gap of 6.2 eV at room temperature (RT) among III-nitrides, with remarkable properties, such as a high breakdown dielectric strength, high chemical and thermal stability, high thermal conductivity, and high melting temperature.

These properties are promising for applications in solid-state light-emitting diodes, radiofrequency power transistors, deep ultraviolet laser diodes, solar-blind photodetectors, and so on [4.16]. Raman scattering has been employed as an important technology in the AlN hot and frontier research [4.18–4.22]. Figure 4.6 presents a typical micro-Raman scattering spectrum of ensemble AlN NWs, showing the AlN characteristic Raman modes of E_2(low), A_1(TO), E_2(high), E_1(TO), and A_1(LO)/E_1(LO) [4.10].

From our investigation "High Quality 10.6 μm AlN Grown on Pyramidal Nano-Patterned Sapphire Substrate by MOCVD" [4.12], temperature-dependent Raman spectroscopy was performed in backscattering geometry $Z(XX)\bar{Z}$ from 80 to 850 K. Figure 4.7(a) shows Raman spectra recorded at typical temperatures of 80, 300, 550, and 850 K, for a 10.6-μm-thick AlN film grown on sapphire, exhibited with the allowed Raman modes of AlN E_2(high) and A_1(LO) modes plus a sapphire mode. Figure 4.7(b) displays the Raman

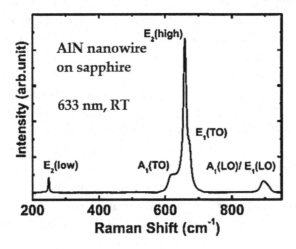

FIGURE 4.6   Micro-Raman scattering spectrum of ensemble AlN nanowires.

Source: From [4.10], figure 1b, with reproduction permission from AIP.

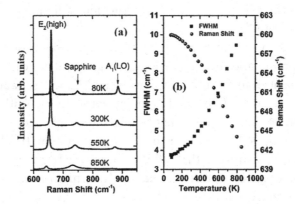

FIGURE 4.7   (a) Raman spectra recorded at 80, 300, 550, and 850 K. (b) Raman linewidth and shift of E_2(high) as a function of temperature.

Source: From [4.12], figure 5, with reproduction permission from AIP.

linewidth and shift of E_2(high) as a function of temperature. The Raman shift at 300 K is determined to be 657.6 cm^{-1} for the AlN epilayer which is in good agreement with the strain-free value of 657.4 ± 0.2 cm^{-1} [4.14, 4.18]. Moreover, the linewidth for the E_2(high) mode of 4.3 cm^{-1} at 300 K was obtained which is comparable to the previously reported value of the AlN bulk single crystal. These results confirm the near strain-free condition of our 10.6-μm-thick AlN epilayer. The decreasing phonon frequencies of the Raman mode are attributed to the thermal expansion of the lattice and phonon-phonon interaction with increasing temperature. The broadening of the Raman linewidth with temperature basically clings to point defect scattering, which reduces the phonon lifetime. All characterization results of Raman measurements together with other techniques have demonstrated that the AlN 10.6-μm-thick epilayer is of high overall crystalline quality [4.12].

In our research "Surface, Structural and Optical Properties of AlN Thin Films Grown on Different Face Sapphire Substrates by Metal-Organic Chemical Vapor Deposition" [4.11], Raman spectroscopy was used to characterize the defect density of AlN thin films grown on different surface orientations, including A– [1120], C– [0001], M– [1010], and R– [1102] faces, of sapphire. Raman line shapes of the AlN E_2 (high) mode for all AlN films were modeled by the spatial correlation model (SCM) (Eqs. (1.16–1.18)), quantifying the MOCVD-grown AlN material properties in correlation with studies of high-resolution X-ray diffraction (HR-XRD), scanning electron microscopy, energy dispersive spectroscopy, and X-ray photoelectron spectroscopy (XPS) [4.11].

In our collaborative investigation "Influence of Dislocations on the Refractive Index of AlN by Nanoscale Strain Field" [4.13], Raman scattering spectra of the five samples with different dislocation densities were exhibited. The Raman shift peak of AlN E_2(high) is found blue-shifted with a decrease in total dislocation density, indicating that the AlN layer suffers more compressive stress from the sapphire substrate. It is found that, with the increase of compressive stress, the refractive index becomes closer to that of bulk AlN. The effect of different dislocation densities on the refractive index of AlN is investigated. With the increase of dislocation densities from 4.24×10^8 to 3.48×10^9 cm^{-2}, the refractive index of AlN decreases from 2.25 to 2.21 at 280 nm. It also demonstrates that the nano-scale strain field around dislocations changes the propagation of light and thus decreases the refractive index of AlN [4.13].

We have performed "A Comparative Study of Multiple Spectroscopies for AlN Thin Films Grown on Sapphire and 6H–SiC by Metal-Organic Chemical Vapor Deposition" [4.14]. Figure 4.8 presents Raman spectra excited by (a) 532nm laser and (b) 266 nm laser for two AlN films grown on 6H–SiC and sapphire, respectively. With the 532 nm excitation, Raman features from the substrate 6H–SiC and sapphire overwhelmed the AlN E_2(high) and A_1(LO) modes. While under the 266 nm excitation, the AlN E_2(high) from the AlN/6H–SiC sample is stronger than all Raman modes from 6H–SiC substrate, and this is so far the case of AlN/sapphire in which AlN E_2(high) is also clearly seen, compatible with other sapphire features. This is because the 532 nm laser line has a large penetrating depth through AlN and under the DUV 266 nm excitation, the laser light penetration depth in AlN is very shallow, which is like the case of 3C–SiC/4H–SiC discussed in Section 2.5.

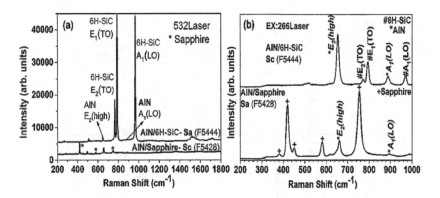

FIGURE 4.8 Raman spectra excited by (a) 532nm laser and (b) 266nm laser for two AlN films grown on 6H–SiC and sapphire, respectively.

Source: From [4.14], figure 5, with reproduction permission from Elsevier.

In the absence of stress, the Raman frequency position of the AlN E_2 (high) phonon mode is 657.4 ± 0.2 cm^{-1} [4.14, 4.18]. From the Gaussian fit of the Raman scattering spectrum with 266 nm excitation for the AlN/sapphire sample-Sa in Figure 4.8(b), the Raman frequency position of the E_2(high) phonon mode can be obtained as 661.3 cm^{-1} – the value is higher by 3.9 cm^{-1} with respect to the case of no stress – caused by compressive stress. For the AlN/6H–SiC sample-Sc, the Raman frequency position of the E_2(high) phonon mode is 654.9 cm^{-1}, with a Raman displacement of -2.5 cm^{-1} relative to the unstressed condition, which is caused by the tensile stress. This is because the thermal expansion coefficient of AlN is greater than 6H–SiC, and the lattice mismatch between the substrate and the epitaxial layer results in tensile stress. The following formula can be used to calculate the stress:

$$\sigma = \Delta\omega / K \tag{4.1}$$

where $\Delta\omega$ is the difference between the E_2(high) phonon peak between the stressed and unstressed AlN epitaxial layer, and K (2.4 ± 0.2 cm^{-1}/GPa) is the strain coefficient [4.14, 4.18].

Figure 4.9 exhibits the 266 nm laser-excited temperature-dependent Raman spectra (TDRS) between 80 K and 800 K for the two A_1(LO) and E_2(high) modes as well as other modes of (a) AlN on 6H–SiC and (b) AlN on sapphire, respectively. The Raman shift of the scattering peak gradually shifts from 80 to 800 K to a low wave number. The Raman shift of the AlN film grown on a sapphire substrate is larger than that on a silicon carbide substrate. Also, the peak width of the Raman scattering peak of the E_2(high) mode gradually widens [4.14]. The stress at each temperature (T) can be obtained by Eq. (4.1).

Figure 4.10 shows the temperature dependence of stress in two AlN films grown on sapphire and 6H–SiC, respectively. It is found that with an increase of T, the stress decreases from positive to negative and that, at low-T, both AlN films have positive, that is, compressive stress, while at high-T, they possess negative, that is, tensile stress. The stress changing

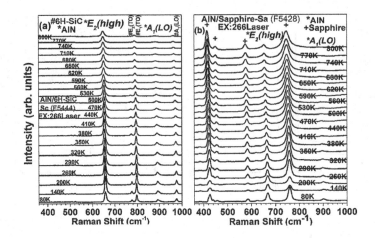

FIGURE 4.9   266 nm laser-excited temperature-dependent Raman spectra of (a) AlN on 6H–SiC and (b) AlN on sapphire, respectively.

Source: From [4.14], figure 6(a) and (b), with reproduction permission from Elsevier.

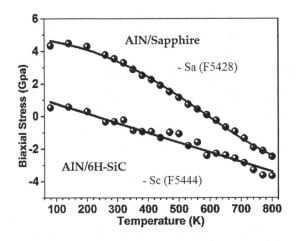

FIGURE 4.10   The temperature dependence of stress in two AlN films grown on sapphire and 6H–SiC, respectively.

Source: From [4.14], figure 8, with reproduction permission from Elsevier.

point of $\sigma = 0$, that is, the biaxial stress changed from compressive stress to tensile stress, is ~200 K for AlN/6H–SiC and near 600 K for AlN/sapphire. This may be caused by the difference between the thermal expansion coefficients of sapphire, 6H–SiC, and aluminum nitride. It is interesting to reveal from TDRS data analyses that the biaxial stress existing in the AlN film changes from compressive to tensile as temperature increases in the range of 80–800 K. The changing point for AlN/sapphire is higher than that of AlN/6H–SiC [4.14].

We have also conducted a Raman investigation of epitaxial AlN thin films on a-, c-, r-face sapphire deposited by the radio-frequency magnetron sputtering [4.15]. Our comprehensive measurement results by Raman scattering together with high-resolution X-ray diffraction (XRD), X-ray photoelectron spectroscopy, spectroscopic ellipsometry, and associated analytical tools clearly show that sapphire substrates of different polarities have effects on the surface roughness, dislocation density, grain size, micro-strain, and surface oxygen binding capacity of the film grown on its surface. From the results of room-temperature Raman characterization, the Raman scattering peak of the AlN film on the C-plane sapphire is obviously weaker than that on the A-plane and R-plane sapphire. In comparison to the three AlN films, the crystallinity of the AlN film grown on the C-plane sapphire is low, and its crystal quality is the worst; the AlN film grown on the A-plane sapphire has the best crystal quality. Variable temperature Raman measurements and analyses have revealed that as the temperature rises from 80 to 800 K, the AlN film has always been tensile stress. The tensile stress of the AlN film grown on the c-plane sapphire has a greater effect on temperature than those grown on a- and r-face sapphire. The lifetime of E_2 (high) phonons gradually decays with the increase in temperature.

Recently, we performed a comprehensive investigation, "Crystalline and Optical Properties of AlN Films with Varying Thicknesses (0.4–10 μm) Grown on Sapphire by Metal-Organic Chemical Vapor Deposition", by way of XRD, optical transmittance, variable angle spectroscopic ellipsometry, and multiple Raman scattering techniques [4.16]. Figure 4.11 presents RT 532 nm excitation Raman spectra of five AlN samples with film thicknesses of 1–10 μm and their E_2(high) mode variations. The E_2(low) at ~249 cm⁻¹, E_2(high) at about 657–659 cm⁻¹, and A_1(LO) at ~887 cm⁻¹ are indicative of Raman-active modes related to c-face AlN. No disorder-activated Raman scattering was observed, indicating AlN films with good crystalline quality. The values of peak frequency and full width at half maximum (FWHM) of the E_2(high) Raman mode of these AlN films are determined accurately, using Gaussian fits.

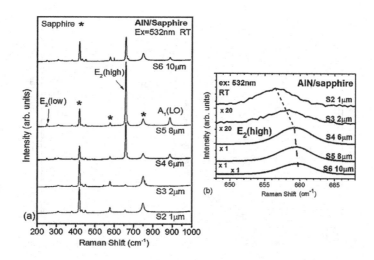

FIGURE 4.11   RT 532nm excitation Raman spectra of five AlN samples: (a) scans between 200–1000 cm⁻¹, (b) the E_2(high) mode.

Source: From [4.16], figure 4(a) and (b), with reproduction permission from Elsevier.

Multiple Raman measurements and analyses were performed with three different excitation wavelengths of visible 532 nm, UV 325 nm, and DUV 266 nm lasers, and in variable temperatures (VT, 80–873 K). The 266-nm light can detect the very top surface of the epitaxial AlN film (about 200–500 nm), and the penetration depth of 325 nm light into AlN is deeper than the 266 nm light. Under the 266 nm excitation, the AlN E_2 (high) mode is observed around 656 cm^{-1} from all samples, less than the stress-free value of 570.4 cm^{-1}, indicating tensile stress near the top surface area. From VT Raman measurements, relationships of both Raman E_2(high) phonon mode frequency and stress (P) for the thickest (10μm) AlN film with temperature (T) were expressed by two polynomial equations. The lifetime of the E_2(high) phonon is longer in low-T than in high-T and decreases with an increase in temperature, due to the thermal expansion of the crystal lattice and the phonon-phonon interaction caused by an increase in temperature. This also causes a change in the residual stress [4.16].

The cross-section Raman scattering measurements along the vertical growth direction were performed on a 10-μm-thick AlN film, with 1 μm step in each detecting point. Figure 4.12 shows RT 532 nm excitation cross-section Raman 12-spectra for the AlN sample S6 (10 μm): (a) scans in 200–1000 cm^{-1}, (b) the A_1(TO), E_2(high), and E_1(TO) three modes. The cross-section Raman measurements revealed the alteration of the E_2(high) mode and stress status. As discussed earlier and in Eq. (4.1), the E_2 (high) peak is sensitive to the stress of the film. The displacement of the E_2 (high) peak from the un-stressed value of 657.4 ± 0.2 cm^{-1} could be used to characterize the stress of films. Moving up from the substrate side, it first shifts toward higher wave numbers and then shifts toward lower wavenumbers. This indicates the changes from compressive to tensile stresses in the AlN film. AlN toward the substrate has more compressive stress and toward the surface has more tensile stress. This change from compressive to tensile stress occurs at about 4 μm

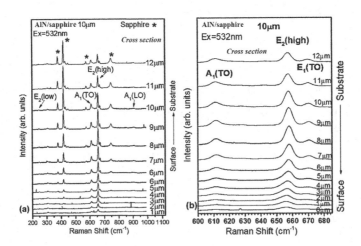

FIGURE 4.12   RT 532nm excitation cross-section Raman spectra for S6 (10 μm): (a) scans in 200–1000 cm^{-1}, (b) the A_1(TO), E_2(high), and E_1(TO) three modes.

Source: From [4.16], figure 7(a) and (b), with reproduction permission from Elsevier.

from the substrate. Also, all six AlN films exhibit compressive stress as whole and tensile stress at the near surface area [4.16].

4.3 RAMAN INVESTIGATION OF M- AND C-FACE AlN CRYSTALS

The structural, surface, and optical properties of m-plane and c-plane AlN crystals were comparatively studied recently by us, using HR-XRD, XPS, and Raman spectroscopy [4.17]. Temperature-dependent Raman measurements show that the Raman shift and the FWHM of E_2 (high) phonon mode of the m-plane AlN crystal were larger than those of the c-plane AlN crystal, which would be correlated with the residual stress and defect in AlN samples, respectively. Meanwhile, the phonon lifetime of the Raman-active modes largely decays, and its line width gradually broadens with the increase in temperature. The phonon lifetime of the Raman TO phonon mode was changed less than that of the LO phonon mode with temperature in the two crystals. It should be noted that the influence of inhomogeneous impurity phonon scattering on phonon lifetime and the contribution to the Raman shift come from the thermal expansion at higher temperatures. In addition, as the temperature increases from 80 to 870 K, the biaxial stress is transformed from compressive to tensile stress.

Figure 4.13 exhibits RT Raman spectra excited by 325 and 532 nm for two bulk m- and c-AlN crystals. It is revealed that using 325/532 nm excitation, m-AlN shows Raman peaks at 247.7 cm^{-1}/247.6 cm^{-1}, 609.2 cm^{-1}/610.4 cm^{-1}, 655.8 cm^{-1}/656.1 cm^{-1}, 668.9 cm^{-1}/669.1 cm^{-1}, and 911.2 cm^{-1}, corresponding to the E_2 (low), A_1 (TO), E_2 (high), E_1 (TO), and E_1 (LO) modes, respectively, which consist of the previously reported data [4.17].

On the other hand, Raman peaks of c-bulk sample at 247.4 cm^{-1}/247.3 cm^{-1}, 655.5 cm^{-1}/656.1 cm^{-1}, and 886.7 cm^{-1}/888.7 cm^{-1} were corresponding to the E_2 (low), E_2 (high), and A_1 (LO) phonon modes of c-plane AlN [4.17], respectively. For c-AlN, it is known that the E_2 (low), E_2 (high), and A_1 (LO) modes are allowed, while the A_1 (TO) and E_1 (TO) phonon modes are prohibited. Temperature-dependent Raman scattering can be employed to probe the temperature effect of epitaxial AlN [4.14–4.16] and bulk AlN crystal

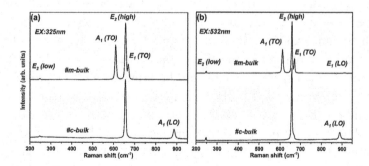

FIGURE 4.13   Room-temperature Raman spectra excited by (a) 325 nm and (b) 532 nm laser for two bulk AlN crystals.

Source: **From [4.17], figure 3, with reproduction permission from the open-access article of MDPI Materials.**

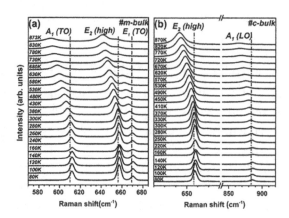

FIGURE 4.14   Temperature-dependent Raman spectra excited by (a) 325 and (b) 532-nm laser for two bulk AlN crystals.

Source: **From [4.17], figure 4, with reproduction permission from the open-access article of MDPI Materials.**

[4.17]. Figure 4.14 shows the variable temperature (80–873/800 K) Raman spectra for bulk AlN crystals excited by the 325 and 532-nm lasers, respectively.

For m-bulk and c-bulk AlN crystals, the frequencies of Raman phonon modes tend to shift toward lower frequencies, and the FWHM values of peaks gradually increase with increasing temperature from 300 to ~870 K, but Raman peaks of two bulk AlN crystals do not have any noticeable change in the lower temperature range (80–300 K). Beyond 300 K, the E_2 (high) peak shifts with temperature are more obvious. For example, peak positions of E_2 (high) of the m-bulk sample are 657.4 cm^{-1} (80 K), 656.2 cm^{-1} (300 K), and 641.3 cm^{-1} (873 K), respectively. Similarly, frequencies of E_2 (high) of the c-bulk sample are 657.5 cm^{-1} (80 K), 656.6 cm^{-1} (300 K), and 643.1 cm^{-1} (870 K), respectively. First, the temperature dependence of Raman frequency shift is related to the change in lattice vibration frequency, caused by either the thermal expansion or contraction of the lattice or the anharmonic effect of lattice vibration resulting from the process of a higher-energy optical phonon decayed into lower energy phonons. Second, the temperature dependence of linewidth could be mainly explained by the phonon–phonon scattering. Finally, the Raman shift of the E_2 mode can reflect the variation of the residual stress in the crystals.

The T-variable Raman spectra of AlN can be fitted with Voigt profile by Eq. (1.28), $\omega(T) = \omega_0 - A / \left\{ \exp\left[B\left(hc\omega_0 / k_B T \right) \right] - 1 \right\}$, which was performed by us for AlN films [4.15] and two m- and c-AlN crystals [4.17]. It has been revealed that the Raman frequency shift of the E_2 (high) phonon mode of m-AlN crystal is affected more significantly than that of c-AlN crystal by temperature. The frequency shift and FWHM of Raman-active modes for two bulk AlN crystals are roughly increased with the increased temperature, while the phonon lifetime is gradually decreased. The Raman frequency shift of the E_2 (high) phonon mode of m-face AlN crystal is easier to affect than that of c-face AlN crystal by temperature. For two AlN crystals, the decay of all the Raman-active modes into two phonons is the prevailing process. We can also obtain information about the impurity-related phonon lifetime from the Raman FWHM. The lifetime of TO phonon mode is less than that of LO phonon mode affected by crystal imperfections, which is attributed to the stronger anharmonic effect on the former. In the higher-temperature region, the relation between Raman shift and temperature is approximately linear, and the phonon frequency shift is strongly influenced by lattice expansion with increasing temperature. Meanwhile, the room-temperature Raman spectra displayed that the FWHM of m-bulk sample is larger than that of c-bulk sample, whereas the residual stress in the m-bulk sample is less than that in the c-bulk sample. In addition, the samples both have inhomogeneous strain, and the stress of the surface layer is larger than that of the inside sample.

4.4 RAMAN INVESTIGATION OF EPITAXIAL InGaN/ GaN ON SAPPHIRE AND ZnO

We have performed Raman scattering studies for epitaxial InGaN/GaN on sapphire [4.23–4.25], including using a UV (325 nm) excitation Raman-photoluminescence microscope [4.23]. For high indium-composition (0.21–0.52) InGaN/GaN heterostructures on ZnO grown by metal-organic chemical vapor deposition (MOCVD), Raman spectra at RT and under 633 nm excitation showed the A_1 (LO) phonon mode varied from 670 down to 610 cm^{-1} [2.24]. For In-rich InGaN layers grown on sapphire by migration-enhanced, plasma-assisted MOCVD, Raman spectra under the excitation of 532-nm laser light, for four $In_xGa_{1-x}N/Al_2O_3$ samples, with In composition x of 0.50, 0.65, 0.75 and 0.80, respectively, showed the E_2(high) and A_1 (LO) phonon modes [4.25].

Figure 4.15 shows Raman spectra excited by 325 nm and measured at RT for an $In_xGa_{1-x}N/GaN$ on sapphire grown by MBE, with x(In) of 0.42. The data were measured by the author, original and unpublished. The InN fraction of the film has been determined using XRD reciprocal space mapping (RSM) [4.26]. The Raman peaks at 540.1, 705.7, and 1392 cm^{-1} are assigned as E_2(high), A_1(LO), and second-order LO of InGaN, respectively,

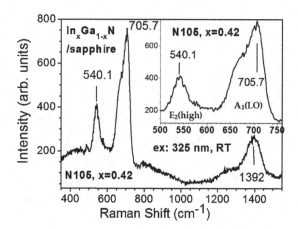

FIGURE 4.15 Raman spectra excited by 325 nm and measured at RT for an $In_xGa_{1-x}N/GaN$ on sapphire grown by MBE, with x(In) of 0.42

Source: Original data and unpublished.

referring to [4.27, 4.28]. Full-compositional $In_xGa_{1-x}N$ films grown by MBE [4.29] and MOCVD [4.30], respectively, are still confronted with serious challenges. We'll make further efforts to perform more Raman measurements and investigation on $In_xGa_{1-x}N$ epitaxial films with large or full composition range.

4.5 RAMAN INVESTIGATION OF EPITAXIAL AlGaN

We have performed Raman scattering studies for epitaxial AlGaN films [4.31–4.34], including RBS investigation of AlGaN with high Al composition [4.31], combined Raman scattering and optical transmission on AlGaN thin films with variable flow rates of trimethylindium [4.32], Raman and material properties of MOCVD-grown AlGaN layers with indium-incorporation [4.33], and an XRD and Raman spectroscopy investigation of AlGaN epilayers with high Al composition [4.34]. In [4.35–4.37], Raman scattering is applied to investigate AlGaN materials.

Three MOCVD-grown AlGaN/AlN/sapphire samples were prepared with variable flow rates of trimethylindium (TMIn), introduced during the growth for the incorporation of indium. They have similar Al compositions (~20%) [4.32]. Figure 4.16 shows the Raman spectra of three AlGaN/AlN/sapphire with different In flow rates (M1710, In = 0 ccm; M1712, In = 50 ccm; M1714, In = 500 ccm), respectively.

Figure 4.17 shows Raman spectra of $Al_xGa_{1-x}N$ epilayers with x = 0.87, 0.58, and 0.15, respectively, measured at RT and under excitation of 325 nm. All the peaks marked with stars are attributed to the phonon modes from the c-plane sapphire substrate. According to the selection rules for Raman scattering, the $Al_xGa_{1-x}N$ epilayer with the hexagonal structure should exhibit phonon of the E_2(high) and A_1(LO) modes [4.34].

The A_1(LO) phonon mode is marked with a vertical arrow in Figure 4.17. Both the E_2(high) and A_1(LO) phonon frequencies are affected by the Al composition in $Al_xGa_{1-x}N$ epilayers. The Raman spectra exhibit intense peaks, without any overtone mode, which also

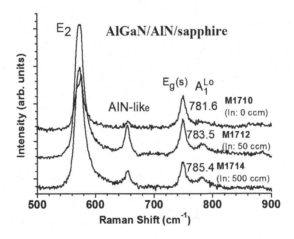

FIGURE 4.16   Raman spectra of three AlGaN/AlN/sapphire with different In flow rates (M1710, In = 0 ccm; M1712, In = 50 ccm; M1714, In = 500 ccm), respectively.

Source: From [4.32], figure 3, with reproduction permission from AIP.

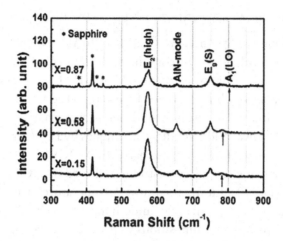

FIGURE 4.17   Raman spectra at room temperature and under excitation of 325 nm of $Al_xGa_{1-x}N$ epilayers with x = 0.87, 0.58, and 0.15, respectively. The peaks marked with stars are attributed to the phonon modes for the sapphire substrate. The vertical arrows show the positions of the $A_1(LO)$ phonon mode forbidden in this scattering geometry.

Source: From [4.34], figure 4, with reproduction permission from Elsevier.

confirm the high quality of the $Al_xGa_{1-x}N$ epilayer samples, with x = 0.87, 0.58, and 0.15, respectively. It indicates that the $E_2(high)$ mode displays a two-mode behavior, showing peaks related to the vibrations of the GaN and AlN sublattices [4.34]. The frequencies of both modes comprised those of pure GaN and AlN and increased with Al composition, allowing the determination of alloy composition. The configurational disorder in the AlGaN epilayer commonly results in asymmetric Raman line shapes for the $E_2(high)$ mode. In an ideal semiconductor, the crystal lattice translation symmetry leads to plane wave phonon eigenstates. Due to the energy and momentum conservation, only q = 0 phonon at the center of the Brillouin zone participates in the Raman scattering process [4.34]. However, configurational disorders or finite-size effects may partially or completely relax the momentum conservation, leading to a downshift and broadening of the Raman phonons mode peak.

However, it can be noticed that with the Al composition gradually increased, the $A_1(LO)$ peaks shift to the higher frequency side obviously in Figure 4.17. The $A_1(LO)$ phonons exhibit a so-called one-mode behavior. The blue shift of the Raman peaks compared with that of GaN is assumed because of phonon confinement or electron-phonon coupling [4.34].

On the other hand, the $E_2(high)$ mode of the AlGaN epilayers shows a little blue shift with an increase in Al composition, which may have arisen from the residual stress. For the AlGaN epilayer grown on sapphire, the large lattice mismatch may result in residual stress in the epilayer. However, the stress due to the lattice mismatch relaxes at a thickness of less than several nanometers near the heterointerface during growth, and, therefore, the residual stress in the epilayer may arise mainly from the large difference in the thermal expansion coefficients between the AlGaN epilayer and sapphire when cooling down from the growth temperature [4.34].

4.6 RAMAN INVESTIGATION OF QUATERNARY InAlGaN

In our collaborative "Raman Spectra Investigation of InAlGaN Quaternary Alloys Grown by Metal-Organic Chemical Vapor Deposition", we presented a Raman scattering investigation of InAlGaN quaternary alloys grown by MOCVD [4.38]. A resonant Raman characterization of InAlGaN/GaN heterostructures was reported by A. Cros et al. [4.39]. The effect of trimethylaluminum flow rate on the AlInGaN quaternary epilayers was studied by Raman scattering and optical and structural techniques by D. Wang et al. [4.40].

Our InAlGaN quaternary films were prepared by MOCVD, first with the AlN buffer layer (25 nm) on the c-plane sapphire substrates, followed by a GaN layer (0.25 μm) and finally the InAlGaN layer (0.23 μm). The growth temperature of 1050°C and pressure of 300 Torr were applied for the growth of the underlying GaN epilayer. InAlGaN layers were grown at 780°C, while In and Al compositions were controlled by varying the flow rates of trimethylindium (TMIn) and trimethylaluminum (TMAl), respectively. The alloy compositions were determined by XPS. Figure 4.18 shows room-temperature (RT) Raman spectra, in the range of 450–800 cm^{-1}, for a series of InAlGaN samples with the composition range of 1.38% < In < 2.73% and 8.01% < Al < 13.97%. Raman peaks from the InAlGaN layer have shown three resolved phonon structures observed in each Raman spectrum of InAlGaN quaternary alloys. In (a)–(c), one peak located in the region of 550–575 cm^{-1}

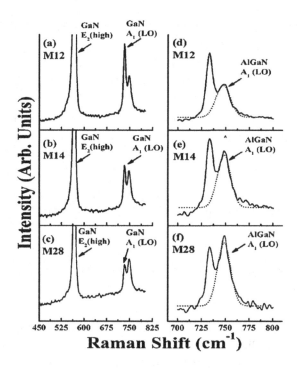

FIGURE 4.18 Room-temperature (RT) Raman spectra of three different InAlGaN samples over the range of 450–825 cm^{-1} (a)–(c), showing GaN E$_2$(high) and GaN A$_1$ (LO) modes at around 735 cm^{-1}, and 700–800 cm^{-1} (d)–(f), showing AlGaN-like A$_1$ (LO) mode at around 750 cm^{-1} described by the Gaussian fit (dotted line) of the Raman line shape.

Source: From [4.38], figure 1, with reproduction permission from AIP.

corresponds to the GaN E_2(high) mode, and the GaN A_1 (LO) mode is located at around 735 cm^{-1}, respectively. In (d)–(f), from 700 to 800 cm^{-1}, the AlGaN-like A_1 (LO) mode is located at around 750 cm^{-1} described by the Gaussian fit (dotted line) of the Raman line shape. The InGaN-like A_1 (LO) mode, expected at around 700 cm^{-1}, is not detectable; probably it is because the In composition is much less than the Al composition in our quaternary samples [4.38].

Figure 4.19 exhibits RT Raman spectra of the AlGaN-like A_1 (LO) and GaN-like A_1 (LO) modes from our quaternary samples M12, M14, and M28 in 675–800 cm^{-1}, respectively. The spectra line shape for all three samples reveals asymmetric Raman linewidth and a shift to the lower frequency of the AlGaN-like A_1 (LO) mode with increasing the ratio of Al/In [4.38].

To fit the spectra line shapes covering both the GaN A_1 (LO) and AlGaN-like A_1 (LO) modes, we modified the SCM (Eq. (1.16)) to involve two items, as shown in eq. (2) of [4.38] and with the fitting results as the solid lines in Figure 4.19. Calculation results indicate that the correlation length L decreases with an increase of the Al/In ratio corresponding to the absence of the long-range order in the alloy. The Raman linewidth of the AlGaN-like A_1 (LO) mode was found to exhibit a maximum at the higher Al/In ratio indicative of a

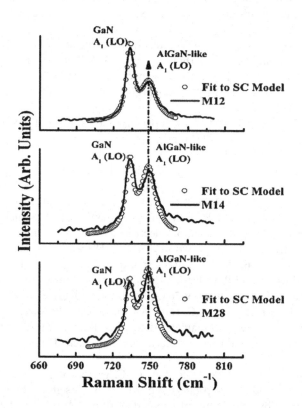

FIGURE 4.19 RT Raman spectra and fits by spatial correlation model (SCM) of the GaN A_1 (LO) and AlGaN-like A_1 (LO) modes from three quaternary samples M12, M14, and M28, in the range of 675–800 cm^{-1}, respectively.

Source: From [4.38], figure 2, with reproduction permission from AIP.

randomly disordered alloy system. The lack of a long-range order exists in the higher Al/In ratio of InAlGaN/GaN alloys. The asymmetric behavior of the AlGaN-like A_1 (LO) Raman linewidth of InAlGaN/GaN heterostructures was attributed to the activation of phonons of q > 0 arising from the disordered state of the alloys [4.38].

REFERENCES

[4.1] C. L. Hsiao, J. T. Chen, H. C. Hsu, Y. C. Liao, P. H. Tseng, Y. T. Chen, Z. C. Feng, L. W. Tu, M. M. C. Chou, L. C. Chen, and K.-H. Chen, "m-plane (10–10) InN heteroepitaxied on (100)-γ-LiAlO2 substrate: Growth orientation control and characterization of structural and optical anisotropy", J. Appl. Phys. **107**, 073502 (2010). https://doi.org/10.1063/1.3359680.

[4.2] Y. L. Chung, X. Peng, Y. C. Liao, S. Yao, L. C. Chen, K. H. Chen, and Z. C. Feng, "Raman scattering and Rutherford backscattering studies on InN films grown by plasma-assisted molecular beam epitaxy", Thin Solid Films **519**, 6778–6782 (2011). https://doi.org/10.1016/j.tsf.2011.04.203.

[4.3] D. N. Talwar, Y. C. Liao, L. C. Chen, K. H. Chen, and Z. C. Feng, "Optical properties of plasma-assisted molecular beam epitaxy grown InN/sapphire", Opt. Mater. **37**, 1–4 (2014). https://doi.org/10.1016/j.optmat.2014.04.012.

[4.4] Z. C. Feng, D. Xie, H.-H. Lin, W. Lu, J.-M. Chen, J. Yiin, K.-H. Chen, L.-C. Chen, B. Klein, and I. T. Ferguson, "Optical, surface and structural studies of InN thin films grown on sapphire by molecular beam epitaxy", J. Vac. Sci. Technol. A **41**, 053401 (2023). https://doi.org/10.1116/6.0002665.

[4.5] V. Y. Davydov, A. A. Klochikhin, V. V. Emtsev, D. A. Kurdyukov, S. V. Ivanov, V. A. Vekshin, F. Bechstedt, J. Furthmüller, J. Aderhold, J. Graul, A. V. Mudryi, H. Harima, A. Hashimoto, A. Yamamoto, and E. E. Haller, "Band gap of hexagonal InN and InGaN alloys", Phys. Stat. Sol. B **234**, 787–792 (2002). https://doi.org/10.1002/1521-3951(200212)234:3<787::AID-PSSB787>3.0.CO;2-H.

[4.6] M. Alevli and N. Gungor, "Effect of N_2/H_2 plasma on the growth of InN thin films on sapphire by hollow-cathode plasma-assisted atomic layer deposition", J. Vac. Sci. Technol. A **38**, 062407 (2020). https://doi.org/10.1116/6.0000494.

[4.7] D. Dobrovolskas, A. Kadys, A. Usikov, T. Malinauskas, K. Badokas, I. Ignatjev, S. Lebedev, A. Lebedev, Y. Makarov, and G. Tamulaitis, "Luminescence of structured InN deposited on graphene interlayer", J. Lumin. **232**, 117878 (2021). https://doi.org/10.1016/j.jlumin.2020.117878.

[4.8] F. M. de Oliveira, A. V. Kuchuk, P. M. Lytvyn, C. Romanitan, H. V. Stanchu, M. D. Teodoro, M. E. Ware, Y. I. Mazur, and G. J. Salamo, "Electron accumulation tuning by surface-to-volume scaling of nanostructured InN grown on GaN (001) for narrow-bandgap optoelectronics", ACS Appl. Nano Mater. **6**, 7582–7592 (2023). https://doi.org/10.1021/acsanm.3c00732.

[4.9] D. Wen, N. Kirkwood, and P. Mulvaney, "Synthesis of size-tunable indium nitride nanocrystals", J. Phys. Chem. Lett. **14**, 3669–3676 (2023). https://doi.org/10.1021/acs.jpclett.3c00024.

[4.10] H. C. Hsu, G. M. Hsu, Y. S. Lai, Z. C. Feng, A. Lundskog, U. Forsberg, E. Janzén, K. H. Chen, and L. C. Chen, "Polarized and diameter-dependent Raman scattering from individual AlN nanowires: The antenna and cavity effects", Appl. Phys. Lett. **101**, 121902 (2012). https://doi.org/10.1063/1.4753798.

[4.11] Y. Li, C. Zhang, X. Luo, Y. Liang, D.-S. Wuu, C.-C. Tin, X. Lu, K. He, L. Wan, and Z. C. Feng, "Surface, structural and optical properties of AlN thin films grown on different face sapphire substrates by metalorganic chemical vapor deposition", Appl. Surf. Sci. **458** (11),972–977 (2018). https://doi.org/10.1016/j.apsusc.2018.07.138.

[4.12] H. Long, J. Dai, Y. Zhang, S. Wang, B. Tan, S. Zhang, L. Xu, M. Shan, Z. C. Feng, H.-C. Kuo, and C. Chen, "High quality 10.6 μm AlN grown on pyramidal nano-patterned sapphire substrate by MOCVD", Appl. Phys. Lett. **114**, 042101 (2019). https://org.doi/10.1063/1.5074177.

[4.13] J. Ben, X. Sun, Y. Jia, K. Jiang, Z. Shi, Y. Wu, C. Kai, Y. Wang, X. Luo, Z. C. Feng, and D. Li, "Influence of dislocations on the refractive index of AlN by nanoscale strain field", Nanoscale Res. Lett. **14**, 184 (2019). https://doi.org/10.1186/s11671-019-3018-7.

[4.14] J. Yin, D. Chen, H. Yang, Y. Liu, D. N. Talwar, T. He, I. T. Ferguson, K. He, L. Wan, and Z. C. Feng, "A comparative study of multiple spectroscopies for AlN thin films grown on sapphire and 6H-SiC by metal organic chemical vapor deposition", J. Alloys Compd. **857**, 157487 (2021). https://doi.org/10.1016/j.jallcom.2020.157487.

[4.15] J. Yin, B. Zhou, L. Li, Y. Liu, W. Guo, D. N. Talwar, K. He, I. T. Ferguson, L. Wan, and Z. C. Feng, "Optical and material properties of polar, semi-polar and non-polar AlN thin films prepared by magnetron sputtering", Semicond. Sci. Tech. **36,** 045012 (2021). https://doi.org/10.1088/1361-6641/abe3c5.

[4.16] Z. C. Feng, H. Yang, L. Wan, F. Wu, J. Dia, C. Chen, V. Saravade, J. Yiin, B. Klein, and I. T. Ferguson, "Crystalline and optical properties of AlN films with varying thicknesses (0.4–10 μm) grown on sapphire by metalorganic chemical vapor deposition", Thin Solid Films **780**, 139939 (2023). https://doi.org/10.1016/j.tsf.2023.139939.

[4.17] S. Zhang, H. Yang, H. Cheng, L. Wang, H. Lu, Y. Yang, L. Wan, G. Xu, I. T. Ferguson, Z. C. Feng, and W. Sun, "Raman spectra and stress investigation of m- and c-face AlN crystals grown by physical vapor transport method", Materials **16**, 1925 (2023). https://doi.org/10.3390/ma16051925.

[4.18] A. Kamarudzaman, A. S. B. Abu Bakar, A. Azman, A.-Z. Omar, A. Supangat, and N. A. Talik, "Positioning of periodic AlN/GaN multilayers: Effect on crystalline quality of a-plane GaN", Mater. Sci. Semicond. Process. **105**, 104700 (2020). https://doi.org/10.1016/j.mssp.2019.104700.

[4.19] S. Hasan, A. Mamun, K. Hussain, M. Gaevski, I. Ahmad, and A. Khan, "Growth evolution of high-quality MOCVD aluminum nitride using nitrogen as carrier gas on the sapphire substrate", J. Mater. Res. **36**, 4360–4369 (2021). https://doi.org/10.1557/s43578-021-00387-z.

[4.20] H. Chang, Z. Liu, S. Yang, Y. Gao, J. Shan, B. Liu, J. Sun, Z. Chen, J. Yan, Z. Liu, J. Wang, P. Gao, J. Li, Z. Liu, and T. Wei, "Graphene-driving strain engineering to enable strain-free epitaxy of AlN film for deep ultraviolet light-emitting diode", Light: Sci. Appl. **11**, 88 (2022). https://doi.org/10.1038/s41377-022-00756-1.

[4.21] Y. Han, H. Zhang, X. Xie, Z. Liu, K. Geng, Y. Yang, Y. Zhang, and G. Chen, "The relationship between annealing and nitrogen flow ratios during magnetron sputtering of AlN films", Appl. Phys. A **129**, 122 (2023). https://doi.org/10.1007/s00339-022-06311-4.

[4.22] J. Gong, J. Zhou, P. Wang, T.-H. Kim, K. Lu, S. Min, R. Singh, M. Sheikhi, H. N. Abbasi, D. Vincent, D. Wang, N. Campbell, T. Grotjohn, M. Rzchowski, J. Kim, E. T. Yu, Z. Mi, and Z. Ma, "Synthesis and characteristics of transferrable single-crystalline AlN nanomembranes", Adv. Electron. Mater. **9**, 220139 (2023). https://doi.org/10.1002/aelm.202201309.

[4.23] Z. C. Feng, M. Schurman, C. Tran, T. Salagaj, B. Karlicek, I. Ferguson, R. A. Stall, C. D. Dyer, K. P. J. Williams, and G. D. Pitt, "Photoluminescence and Raman scattering characterization of GaN, InGaN and AlGaN films using a UV excitation Raman-photoluminescence microscope", Mater. Sci. Forum **264–268**, 1359–1362 (1998). https://doi.org/10.4028/www.scientific.net/MSF.264-268.1359.

[4.24] M. Tian, Y. Qian, C. Zhang, L. Li, S. Yao, I. T. Ferguson, D. N. Talwar, J. Zhai, D. Meng, K. He, L. Wan, and Z. C. Feng, "Investigation of high indium-composition InGaN/GaN heterostructures on ZnO grown by metallic organic chemical vapor deposition", Opt. Mater. Exp. **8** (10), 3184 (2018). https://doi.org/10.1364/OME.8.003184.

[4.25] C. Zhang, Y. Li, Y. Liang, D. N. Talwar, S.-Y. Huang, Q. Li, F. Wang, L. Wan, D. Seidlitz, N. Dietz, D.-S. Wuu, K. He, and Z. C. Feng, "Surface and optical properties of In-rich InGaN layers grown on sapphire by migration-enhanced plasma assisted metal organic chemical vapor deposition", Mater. Res. Exp. **6**, 016407 (2019.01). https://doi.org/10.1088/2053-1591/aae4b5.

[4.26] S. T. Liu, X. Q. Wang, G. Chen, Y. W. Zhang, L. Feng, C. C. Huang, F. J. Xu, N. Tang, L. W. Sang, M. Sumiya, and B. Shen, "Temperature-controlled epitaxy of $In_xGa_{1-x}N$ alloys and their band gap bowing", J. Appl. Phys. **110**, 113514 (2011). http://dx.doi.org/10.1063/1.3668111.

[4.27] R. Oliva, J. Ibanez, R. Cusco, R. Kudrawiec, J. Serafinczuk, O. Martınez, J. Jimenez, M. Henini, C. Boney, A. Bensaoula, and L. Artús, "Raman scattering by the E2h and A1(LO) phonons of $In_xGa_{1-x}N$ epilayers (0.25 < x < 0.75) grown by molecular beam epitaxy", J. Appl. Phys. **111**, 063502 (2012). http://dx.doi.org/10.1063/1.3693579.

[4.28] R. Chai, L. Wang, L. Wen, W. Li, S. Zhang, W. Wei, W. Sun, and S. Yang, "Raman spectra of semi-polar (11–22) InGaN thick films", Vib. Spectrosc. **119**, 103357 (2022). https://doi.org/10.1016/j.vibspec.2022.103357.

[4.29] X. T. Zheng, T. Wang, P. Wang, X. X. Sun, D. Wang, Z. Y. Chen, P. Quach, Y. X. Wang, X. L. Yang, F. J. Xu, Z. X. Qin, T. J. Yu, W. K. Ge, B. Shen, and X. Q. Wang, "Full-composition-graded $In_xGa_{1-x}N$ films grown by molecular beam epitaxy", Appl. Phys. Lett. **117**, 182101 (2020). https://doi.org/10.1063/5.0021811.

[4.30] J. Zhao, K. Chen, M. Gong, W. Hu, B. Liu, T. Tao, Y. Yan, Z. Xie, Y. Li, J. Chang, X. Wang, Q. Cui, C. Xu, R. Zhang, and Y. Zheng, "Epitaxial growth and characteristics of nonpolar a-plane InGaN films with blue-green-red emission and entire in content range", Chin. Phys. Lett. **39**, 048101 (2022). https://doi.org/10.1088/0256-307X/39/4/048101.

[4.31] T. Y. Lin, Y. L. Chung, L. Li, S. Yao, Y. C. Lee, Z. C. Feng, I. T. Ferguson, and W. Lu, "Optical, structural and nuclear scientific studies of AlGaN with high Al composition", SPIE Proc. **7784**, 778415 (2010). https://doi.org/10.1117/12.859106.

[4.32] Y. L. Chung, T. Y. Lin, Z. C. Feng, **D. S. Wuu,** I. Ferguson, and W. Lu, "Combined Raman scattering and optical transmission on AlGaN thin films with variable flow rates of trimethylindium (International Conference on Raman Spectroscopy (ICORS) 2010)", AIP Conf. Proc. **1267**, 1113–114 (2010). https://doi.org/10.1063/1.3482326.

[4.33] T.-Y. Lin, L. Li, C. Chen, Y.-L. Chung, S. D. Yao, Y.-C. Lee, Z. R. Qiu, I. T. Ferguson, D.-S. Wuu, and Z.C. Feng, "Material properties of MOCVD grown AlGaN layers influenced by indium-incorporation", Proc. SPIE **8123**, 812309 (2011), https://doi.org/10.1117/12.892932.

[4.34] S. Wang, X. Zhang, Q. Dai, W. Guo, F. Li, J. Feng, Z. C. Feng, and Y. Cui, "An x-ray diffraction and Raman spectroscopy investigation of AlGaN epi-layers with high Al composition", Optik **131**, 201–206 (2017). https://doi.org/10.1016/j.ijleo.2016.11.079.

[4.35] Y. Liu, X. Yang, D. Chen, H. Lu, R. Zhang, and Y. Zheng, "Determination of temperature-dependent stress state in thin AlGaN layer of AlGaN/GaN HEMT heterostructures by near-resonant Raman scattering", Adv. Condens. Matter Phys. **2015**, 918428 (2015). http://dx.doi.org/10.1155/2015/918428.

[4.36] D. Kosemura, V. Sodan, and I. D. Wolf, "Correlation between temperature dependence of Raman shifts and in-plane strains in an AlGaN/GaN stack", J. Appl. Phys. **121**, 035702 (2017). http://dx.doi.org/10.1063/1.4974366.

[4.37] Y. Liu, D. Chen, G. Wei, Z. Lin, A. He, M. Li, P. Wang, R. Zhang, and Y. Zheng, "Temperature-dependent ultraviolet Raman scattering and anomalous Raman phenomenon of AlGaN/GaN heterostructure", **Opt. Exp. 27**, 4781–4788 (2019). https://doi.org/10.1364/OE.27.004781.

[4.38] S. Y. Hu, Y. C. Lee, Z. C. Feng, and Y. H. Weng, "Raman spectra investigation of InAlGaN quaternary alloys grown by metalorganic chemical vapor deposition", J. Appl. Phys. **112**, 063111 (2012). https://doi.org/10.1063/1.4752420.

[4.39] A. Cros, A. Cantarero, N. T. Pelekanos, A. Georgakilas, J. Pomeroy, and M. Kuball, "Resonant Raman characterization of InAlGaN/GaN heterostructures", Phys. Stat. Sol. B **243**, 1674–1678 (2006). https://doi.org/10.1002/pssb.200565132.

[4.40] D. Wang, G. Liu, S. Jiao, L. Kong, T. Liu, T. Liu, J. Wang, F. Guo, C. Luan, and Z. Li, "The effect of trimethylaluminum flow rate on the structure and optical properties of AlInGaN quaternary epilayers", Crystals 7, 69 (2017). https://doi.org/10.3390/cryst7030069.

CHAPTER 5

III-V Semiconductors

5.1 RAMAN SCATTERING OF InSb

InSb has the highest electron mobility and narrowest band gap among III-V compound semiconductors and is attractive for applications in high-speed electronic and optoelectronic devices in the infrared. Various growth techniques have been applied to grow InSb on GaAs, including molecular beam epitaxy (MBE), liquid phase epitaxy, metal-organic magnetron sputtering (MOMS), or metal-organic chemical vapor deposition (MOCVD) [5.1–5.3]. We have performed Raman scattering investigation for InSb films grown on GaAs by MOMS [5.1], resonance Raman scattering from epitaxial InSb thin films [5.2], and Raman characterization and mapping of four-inch InSb epitaxial thin films grown on GaAs by MOCVD [5.3]. According to the literature, Raman scattering has been reported as a powerful method in the investigation of InSb and related materials [5.4–5.7].

Figure 5.1 shows Raman spectra at 80 K and under excitation at 476, 488, and 496 nm, for MOMS-grown InSb film on GaAs (left) and intrinsic bulk lnSb (right), respectively [5.2]. The enhancement of the InSb 2LO phonon band for InSb/GaAs and especially for bulk InSb is seen from outgoing resonant Raman scattering, under 488 nm (2.539 eV) excitation, where the energy of the Raman-scattered photon matches a gap energy of the InSb $E_1+\Delta_1$ gap (2.491 eV). InSb has the 1LO energy of 23.9 meV, and the 2.491 eV adds 2 × 0.0239 eV reaching 2.539 eV, matching the energy of incident 488 nm laser energy.

Figure 5.2 exhibits Raman spectra, measured at 80 K and the 488 nm excitation, of five MOMS-grown InSb/GaAs samples with different thicknesses d(InSb) of 0.17–1.67 μm [5.2]. All Raman 1LO and 2LO peaks have a full width at half maximum (FWHM) < 6 cm^{-1}, indicating good crystalline structure, despite a 14.6% lattice mismatch between InSb and GaAs.

The FWHMs, the ratio of intensities I(2LO)/I(1LO), and the longitudinal optical (LO) peak frequencies vs d(InSb) were studied [5.2]. It is found that the FWHM of the LO line decreases as d(InSb) increases from 0.17 to 0.9 μm, reaching a minimum of 4 cm^{-1}, indicating that film structural quality improves with d(lnSb). The subsequent apparently anomalous increase in FWHM to 5 cm^{-1} for the thickest (1.67 μm) film was ascribed to the relatively poor quality of this film. The intensity ratio I(2LO)/I(1LO) was found to decrease

DOI: 10.1201/9781032644912-5

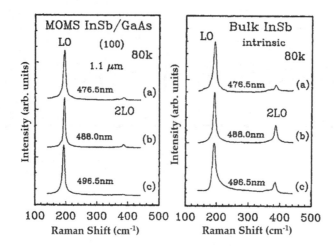

FIGURE 5.1   Raman spectra at 80 K and under excitation at 476, 488, and 496 nm, for MOMS-grown InSb film on GaAs (left) and intrinsic bulk InSb (right).

Source: From [5.2], figures 2 and 3, with reproduction permission of AIP.

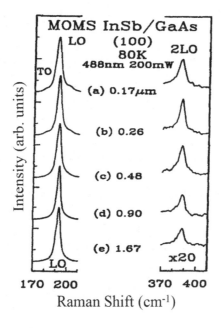

FIGURE 5.2   Raman spectra of five MOMS-grown InSb/GaAs samples with different film thicknesses of 0.17–1.67 μm.

Source: From [5.2], figure 3, with reproduction permission of AIP.

with d(InSb). Also, a weak feature appears on either side of the LO phonon band for the two thinnest films, in Figure 5.2. The 182 cm^{-1} feature lies at the nominally forbidden transverse optical (TO) phonon frequency for lnSb and reflects the partly disordered structure of very thin films on a substrate with a large mismatch. The higher feature, near 200 cm^{-1}, may come from dislocations induced by the lattice mismatch. Both features weaken as d(InSb) increases. We speculate that this behavior might arise from the fundamental growth process. Microphotographs show that, in the early growth, the deposited InSb consists of partially connected islands growing in all three dimensions. Island growth continues until the 1nSb layer becomes continuous beyond 0.5 μm, with all the islands precisely aligned with each other to form an epitaxial layer. Thin layers can relax when cooled from growth, since the separate islands accommodate thermal changes by bulging out. Thick continuous layers, however, lack this accommodation, so increased compressive stress appears with increasing thickness [5.2].

5.2 RAMAN SCATTERING OF GaSb

GaSb has an energy bandgap of 0.70 eV (1.77 μm) at room temperature (RT) and 0.81 eV (1.53 μm) at 4 K. It is an important material for infrared (IR) optoelectronic and electronic devices in the wavelength range of 1–5 μm [5.8]. We have conducted "Optical Investigation of GaSb Thin Films Grown on GaAs by Metalorganic Magnetron Sputtering" [5.8]. Raman scattering studies on GaSb were also performed in [5.9, 5.10]. Figure 5.3 shows Raman spectra from three MOMS-grown GaSb/GaAs (001) grown at substrate temperatures of (a) 480°C, (b) 440°C, and (c) 400°C, and (d) a bulk GaSb [5.8]. The GaSb LO and TO phonon modes at the center of the Brillouin zone were detected.

According to Raman selection rules, the LO phonon mode is allowed and TO is forbidden for the (001) orientation of a zincblende crystal. Because these GaSb films are thinner than 0.6 μm, there exists a high density of dislocations in the films, which breaks the symmetry and leads to the appearance of the forbidden TO modes. As seen in Figure 5.3(a),

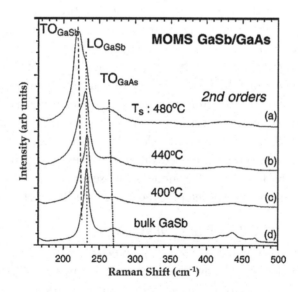

FIGURE 5.3   Raman spectra from three MOMS-grown GaSb/GaAs (001) grown at substrate temperatures of (a) 480°C, (b) 440°C, and (c) 400°C and (d) a bulk GaSb, respectively.

Source: From [5.8], figure 1, with reproduction permission of Elsevier.

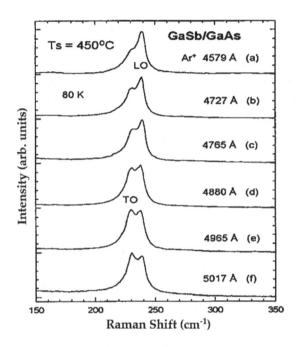

FIGURE 5.4   Raman spectra from a MOMS-grown GaSb on GaAs (100) with Ts = 450°C, under different excitation wavelengths between 4579 and 5017 Å from an Ar+ laser.

Source: From [5.8], figure 3, with reproduction permission of Elsevier.

the "forbidden" TO mode is dominant for samples grown at high substrate growth temperature Ts = 480°C. As Ts decreases to 440°C, in Figure 5.3(b), the LO mode intensity increases with respect to the TO mode. At Ts = 400°C, the allowed LO mode is dominant with a weak shoulder, "forbidden" TO mode, in Figure 5.3(c). This indicates that the film quality improves with decreasing substrate growth temperature, which is consistent with the X-ray diffraction (XRD) result showing a narrowing of the (004) GaSb peak as decreasing Ts (data not shown here and in [5.8]). The optimum substrate temperature for the growth of epitaxial GaSb thin films using the MOMS technique is about 400°C [5.8].

Figure 5.4 exhibits Raman spectra from a MOMS-grown GaSb on GaAs (100) with Ts = 450°C, under different excitation wavelengths between 4579 and 5017 Å from an Ar+ laser [5.8]. As seen, with decreasing excitation wavelength or penetration depth from Figure 5.4(f) to (a), the forbidden TO mode decreases in intensity with respect to the allowed LO mode, indicating the development of the crystalline perfect toward the surface.

5.3 RAMAN SCATTERING OF Si-DOPED InAs

Devki N. Talwar, Hao-Hsiung Lin, and Zhe Chuan Feng conducted the following research: "Phonon Characteristics of Heavily Si-Doped InAs Grown by Gas Source Molecular Beam Epitaxy" [5.11]. There are more studies of Raman scattering on InAs [5.12–5.14]. The experimental Si-doped InAs films were grown on an n-type InAs (100) substrate by a VG-V80H gas-source molecular beam epitaxy (GS-MBE) system. Raman spectroscopic measurements were performed using a T64000 Jobin Yvon triple Raman spectrometer by

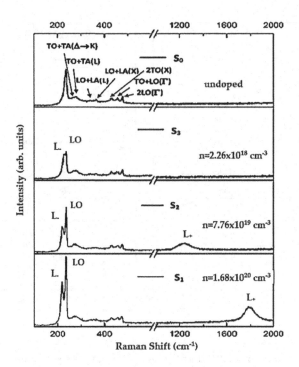

FIGURE 5.5   Raman scattering spectra recorded in near-backscattering geometry under 514-nm excitation. The results for an undoped InAs sample, S_0, and three Si-doped InAs samples S_1, S_2, and S_3 with different charge carrier concentrations n are displayed, respectively. The two-phonon modes are observed in all samples, marked above the spectrum of the S_0 sample.

Source: **From [5.11], figure 2, with reproduction permission of Wiley.**

employing Ar-ion 514 nm and He-Ne 633 nm of laser light sources, respectively. The power of the laser beam and its size in diameter are kept at 10 mW and ~10 µm, respectively, to reduce the effects of laser-induced heating of the samples and to minimize the generation of photoexcited charge carriers. The scattered light is analyzed by using a spectrometer equipped with a charge-coupled device. In the Raman measurements, we kept the spectral resolution of the spectrometer at ~1 cm^{-1}.

Figure 5.5 presents Raman scattering spectra of three Si-doped InAs samples, recorded in near-backscattering geometry at RT and with a 514-nm excitation. The Raman spectroscopic data for an undoped InAs sample, S_0, and three Si-doped InAs samples S_1, S_2, and S_3 with different charge carrier concentrations n are displayed, respectively. The two-phonon modes are observed in all samples and assigned in the S_0 sample only by incorporating group theoretic selection rules and critical point energies derived from the rigid ion model [5.11].

5.4 RAMAN SCATTERING OF AlGaAs

With its fundamental interest and importance in optoelectronics, $Al_xGa_{1-x}As$ has been a widely studied ternary compound semiconductor for many years. The optical phonons of this system are of special significance. We have performed a Raman scattering

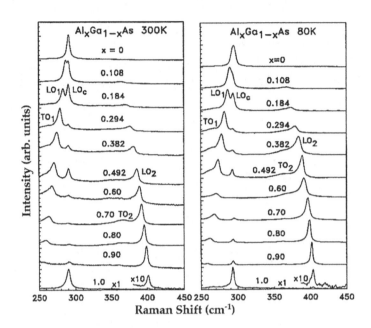

FIGURE 5.6 Raman spectra of $Al_xGa_{1-x}As$ over the entire x range, measured at 300 K (left) and 80 K (right), excited at 488 nm and 100 mW power. x = 0, GaAs substrate material; x = 1, AlAs/GaAs superlattice. Other samples, MBE-grown $Al_xGa_{1-x}As$/GaAs films. TO, transverse optical mode; LO, longitudinal optical mode; subscripts 1 and 2, GaAs-like and AlAs-like, LO_c represents the LO phonon from the thin GaAs cap layer, respectively.

Source: From [5.16], figures 1 and 2, with reproduction permission of APS.

investigation for AlGaAs [5.15–5.17]. There are other Raman studies on AlGaAs in the literature [5.18–5.20]. Figure 5.6 shows Raman spectra of $Al_xGa_{1-x}As$ over the entire x range, $0 \leq x \leq 1$, measured at 300 K (left) and 80 K (right), excited at 488 nm and 100 mW power, where x = 0, that is, the GaAs substrate material [5.16].

A full set of 11 $Al_xGa_{1-x}As$ samples with aluminum (Al) composition x = 0–1 in steps of 0.1 was studied. Nine $Al_xGa_{1-x}As$ films with x = 0.1–0.9 were made at the Lockheed Palo Alto Research Laboratories by MBE. The 1-μm-thick films were grown on semi-insulating (100) GaAs wafers, held at 580–620°C. Each film was capped with an approximately 20-nm GaAs passivation layer. The films were examined by reflection high-energy electron diffraction, Auger, and x-ray methods, which showed excellent crystallinity and uniformity. A piece of the commercial pure GaAs substrate material gave our data for x = 0. We obtained data on AlAs as they appeared in a superlattice with 167 pairs of AlAs and GaAs layers, each 15 nm thick, grown on GaAs. Raman spectra were measured in a near-backscattering geometry and excited by the 488-nm line from an Ar+ laser operating at 100 mW. A triple spectrometer SPEX 1877 with a cooled optical multichannel analyzer was employed for detection, providing superior signal-to-noise ratios and a resolution of 1–2 cm^{-1} [5.16].

The frequencies of the LO peaks in Figure 5.6 are plotted vs x in Figure 5.7, at 300 and 80 K, respectively. A modified random element isodisplacement (MREI) model was applied

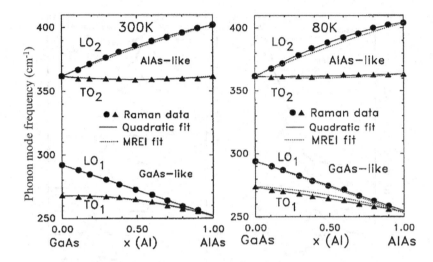

FIGURE 5.7 Longitudinal- and transverse-optical-phonon frequencies vs x in $Al_xGa_{1-x}As$ at 300 K (left) and at 80 K (right). Fits from a modified random element isodisplacement (MREI) model were applied, showing Raman data, quadratic least-squares fits from Eqs. (5.1 and 5.2).

Source: From [5.16], figures 3 and 4, with reproduction permission of APS.

to fit the experimental Raman LOs and TOs frequencies, and we have obtained good fits within the experimental accuracy with quadratic or linear expressions. At 300 K, the best fits in the least-squares sense are [5.16]:

$$\omega(LO_1) = 292.1 - 39.96x, \tag{5.1}$$

$$\omega(LO_2) = 361.7 + 55.62x - 15.45x^2 \tag{5.2}$$

whereas the best fits to the data at 80 K are

$$\omega(LO_1) = 294.0 - 38.95x - 0.45x^2 \tag{5.3}$$

$$\omega(LO_1) = 361.5 + 65.61x - 22.68x^2 \tag{5.4}$$

These are represented by solid lines in Figure 5.7, respectively.

Following the previous 300 and 80 K Raman spectral data of $Al_xGa_{1-x}As$ over full x = 0–1, we employed the spatial correlation model (SCM) (Eqs. (1.16–1.18)) to further analyze the data in "Characterization of MBE-Grown $Al_xGa_{1-x}As$ Alloy Films by Raman Scattering" [5.17], as shown in Figure 5.8.

The obtained reverse of correlation length L as a function of composition x, that is, 1/L vs x (Al or Ga), is presented in Figure 5.9. It is found that the correlation length mainly depends on the composition and that no significant temperature effect was found. The comparison of GaAs-like and AlAs-like modes indicates that the correlation length is also slightly dependent on the phonon mode itself. A more plausible mechanism to account for this disorder effect is the random distribution of compositions in alloys.

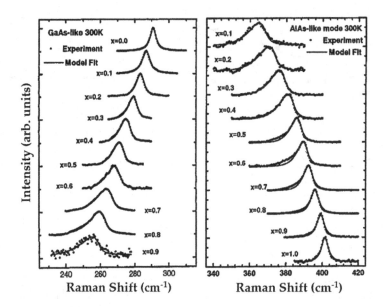

FIGURE 5.8 Room-temperature Raman spectra and SCM fits of $Al_xGa_{1-x}As$ films for different Al compositions in GaAs-like LO (left) and AlAs-like LO (right) phonon frequency region.

Source: From [5.17], figures 1 and 2, with reproduction permission of Wiley.

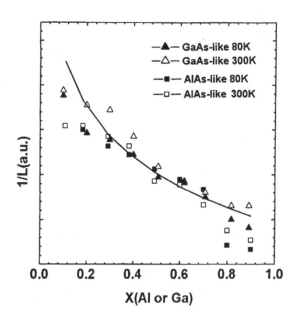

FIGURE 5.9 The reverse of correlation length L as a function of composition x. Composition x is the composition of Ga for the GaAs-like mode and Al for the AlAs-like mode. The solid line indicates the configurational entropy of mixing obtained for an ideal alloy (in arbitrary units).

Source: From [5.17], figure 4, with reproduction permission of Wiley.

5.5 RAMAN SCATTERING OF InGaAs

The $InxGal_{1-x}As$/InP material system is of considerable interest for laser diodes and photo-detectors in long-wavelength optical communications and is promising for high-frequency field-effect transistors and heterojunction photovoltaic cells. We have performed a Raman investigation on InGaAs grown on InP by radial flow epitaxy (URFE) [5.21]. Other Raman works on InGaAs can be found in [5.22–5.24].

Figure 5.10 shows Raman spectra at 300 and 80 K with 100 mW excitation at 5145 Å, from three $In_xGa_{1-x}As$/InP (0.52 < x < 0.54) with thicknesses of top $In_xGa_{1-x}As$ layers of (a) 1.43 μm, (b) 95 Å, and (c) 20 Å, in both [I]-left graph for 300 K data and [II]-right graph for 80 K data, respectively. The penetration depth at wavelengthλ is defined by $\delta_p = 1/\alpha = \lambda/4\pi k$, where α is the absorption coefficient, and k is the extinction coefficient. Using reported values of k for InP, InAs, and GaAs (0.37–1.3 at 2.40 eV), the penetration depth at 5145 Å is among 300–1100 Å. Assuming δ_p for InGaAs is between that of InAs and GaAs, the Raman signals collected must have originated from the first 1100 Å of each sample. Thus, for the thick 1.43 μm layer, the Raman signal came from the InGaAs region only without features from the InP substrate. For both the 95 and 20 Å thin InGaAs films, the Raman signals from the top InGaAs layer and the underlying InP layer were observed together [5.21].

We observed Raman lines from the underlying InP layer in the 95 and 20 Å InGaAs/InP structures (Figure 5.10[I]–[II](b) and (c)). There are the first-order LO and TO phonons, LO_p and TO_p from InP, and their second-order overtones and combinations (2LO, $2TO_p$, and LO_p + TO), respectively. Raman features between 200 and 300 cm⁻¹ in Figure 5.10[I]

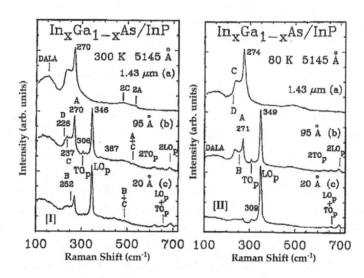

FIGURE 5.10   Raman spectra from three $In_xGa_{1-x}As$/InP (0.52 < x < 0.54) at 300 K (left) and 80 K (right) with 100 mW excitation at 5145 Å. The thicknesses of the top $In_xGa_{1-x}As$ layers are (a) 1.43 μm, (b) 95 Å, and (c) 20 Å, in both graphs of [I]-left and [II]-right, respectively.

Source: From [5.21], figures 1 and 2, with reproduction permission from AIP.

and [II] graphs came from two modes of the InGaAs alloy. The GaAs-like LO mode, labeled A near 270 cm^{-1}, is the major band. The InAs-like LO mode, labeled C around 237 cm^{-1}, is also strong. The GaAs-like TO mode, B near 252 cm^{-1}, is located between A and C, and the InAs-like TO mode, D near 226 cm^{-1}, appears as the low wavenumber shoulder of C.

The intensity of InP LO$_p$ relative to InGaAs alloy modes increases with a decrease of the InGaAs film thickness, which reflects the variation of excitation penetrating the InP sublayer. Second-order alloy phonons appear as weak bands between 450 and 550 cm^{-1}. Broad features below 200 cm^{-1} represent the disorder-activated longitudinal acoustic phonon and other defect-related features. The dominant and sharp features of the InP LO$_p$ and GaAs-like LO characterize the good quality of InP substrate and InGaAs epitaxial layers.

The dependence of Raman line shifts on temperature for thick and thin InGaAs films leads to an analysis of the layer stress and strain. We found that GaAs-like LO phonon line A has shifted from 270 cm^{-1} at 300 K (Figure 5.10 [I]-(a)) to 274 cm^{-1} at 80 K (Figure 5.10 [II]-(a)) for the 1.43 μm thick film. However, for the 95 Å thin film, line A of 270 cm^{-1} at 300 K (Figure 5.10 [I]-(b)) has shifted only to 271 cm^{-1} at 80 K (Figure 5.10 [II]-(b)). At the growth temperature (650°C), In$_{0.53}$Ga$_{0.47}$As has a lattice constant larger than that of InP and the layer is under compression. At RT, In$_{0.53}$Ga$_{0.47}$As/InP is almost perfectly lattice matched. If we further cool to 80 K, the In$_{0.53}$Ga$_{0.47}$As lattice constant becomes less than that of InP. For the thick film in Figure 5.10[I]-(a) and [II]-(a), the InGaAs layer stress was released by misfit dislocations, and its lattice constant varied mainly with temperature, especially for the near-surface region detected (~1100 Å), which is far from InP/InGaAs interface [5.21].

5.6 RAMAN SCATTERING OF InGaSb

In$_{1-x}$Ga$_x$Sb, a III-V semiconducting alloy, is a promising material for infrared optical devices. Its lattice constant varies linearly with composition from 6.48 Å at x = 0 to 6.10 Å at x = 1. Its direct energy gap ranges monotonically from 0.17 eV in InSb to 0.72 eV in GaSb, which makes it a good material for infrared detectors operating over 1.5–5 μm [5.25]. We have performed a Raman scattering investigation on InGaSb [5.25, 5.26]. There exist some more Raman studies for InGaSb and related materials [5.27–5.30]. Figure 5.11 presents Raman spectra for a series of In$_{1-x}$Ga$_x$Sb grown on GaAs with different x(Ga) by way of the MOMS technology [5.25]. We used Raman scattering and infrared-reflectivity measurements to examine In$_{1-x}$Ga$_x$Sb films of thickness 1–2 μm with x(Ga) = 0.0, 0.035, 0.07, 0.24, 0.32, 0.68, and 1. For x = 0, that is, pure InSb on GaAs, in Figure 5.11 (a), only InSb phonon features appear, including the LO, TO modes, and the 2LO. For x = 1, that is, pure GaSb on GaAs, in Figure 5.11(g), the GaSb LO mode at 238 cm^{-1} and TO mode at 228 cm^{-1} appeared as the strong shoulder of the LO band. The two-mode phonon behavior over the range 0 < x(Ga) < 0.7 is observed. For In$_{1-x}$Ga$_x$Sb with small x values, InSb-like features dominate the spectra, as shown in Figure 5.11(b) and (c). Both InSb-like and GaSb-like phonons are clearly seen for x = 0.24 and 0.32 in Figure 5.11(d) and (e), and weakly for x = 0.68 in Figure 5.11(f). The Raman line widths at half maximum are < 10 cm^{-1} and the polarization behavior of the Raman spectra obeys the selection rules for a (001) surface, indicating that the films are of good crystalline quality.

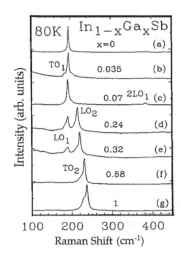

FIGURE 5.11 Raman spectra for MOMS-grown $In_{1-x}Ga_xSb$ on GaAs with x(Ga) = (a) 0 (pure InSb on GaAs), (b) 0.035, (c) 0.07, (d) 0.24, (e) 0.32, (f) 0.68, and (g) 1 (pure GaSb on GaAs), respectively, measured at 80 K and under excitation of 501.7 nm and 100 mW from an Ar^+ laser.

Source: From [5.25], figure 1, with reproduction permission from Can. Sci. Pub.

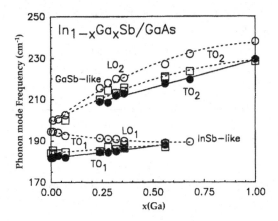

FIGURE 5.12 Relationship of TO and LO phonon mode frequencies vs x(Ga) for epitaxial $In_{1-x}Ga_xSb$ on GaAs. Data at 300 K and at 80 K are represented by filled symbols and open symbols, respectively. Solid lines and dashed lines represent second-order regression fits to the room temperature and nitrogen-temperature data, respectively. The LO mode frequencies are extracted from Raman spectra whereas the TO mode frequencies are extracted from the fits to the infrared data.

Source: From [5.26], figure 5, with reproduction permission from APS.

Infrared-reflectivity (IR) spectral measurements were performed for this set of $In_{1-x}Ga_xSb$ samples at 300 K [5.25] and 80 K [5.26]. The IR spectra over 50–450 cm^{-1} and analyses yield phonon mode and transport information for these samples. The dependence of the GaSb- and InSb-like TO and LO phonon frequencies on Ga composition can be derived, as shown in Figure 5.12.

At 300 K, the TO phone frequencies of $In_{1-x}Ga_xSb$ are linear with x(Ga), and can be expressed as [5.26]:

$$\omega_{TO}\left(InSb-like\right) = 181.7 + 11.2x,\ 0 \le x \le 0.56, \tag{5.5}$$

$$\omega_{TO}\left(GaSb-like\right) = 202.4 + 26.7x,\ 0.24 \le x \le 1. \tag{5.6}$$

These equations provide a simple way to determine the alloy composition from RT IR data.

5.7 RAMAN SCATTERING OF QUATERNARY InGaAlAs

The $In_{1-x-y}Ga_xAl_yAs$ materials are attractive for opto-electronics because of their suitable bandgap range (0.76–1.5 eV), large bandgap discontinuity, and high electron saturation velocity. These III-III-III-V quaternaries consist of only one group V anion of As, which is much less volatile than phosphorous, and the group III elements of In, Ga, and Al possess almost unity sticking coefficients, which makes them very suitable for epilayer growth by MBE [5.31]. Similar studies, including Raman scattering, on InGaAlAs were conducted also in [5.32, 5.33]. InP has been widely adopted as the substrate for the growth of these quaternary alloys, which are lattice matched with InP for x+y = 0.47. We have prepared a series of $In_{1-x-y}Ga_xAl_yAs$ thin films on InP, grown by MBE under different conditions [5.31]. A comprehensive characterization was performed by DCXRD, photoluminescence (PL), Fourier transform infrared (FTIR), and Raman techniques on these structures.

Figure 5.13 exhibits Raman spectra from three samples with different x and y. Three mode behaviors are seen. For sample #298 with a high x = 58.6% and a low In composition

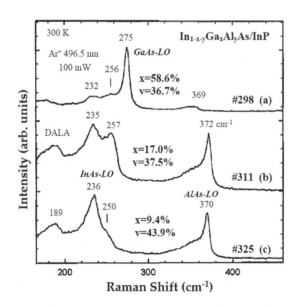

FIGURE 5.13   Raman spectra of three $In_{1-x-y}Ga_xAl_yAs$/InP with different compositions, measured at 300 K and under excitation of 4965 Å and 100 mW.

Source: From [5.31], figure 2, with reproduction permission from IEEE.

(4.7%), the GaAs-like LO phonon mode at 275 cm^{-1} is dominant in Figure 5.13(a). As x decreases and y increases, the InAs-like LO at 235 cm^{-1} and AlAs-like LO at 372 cm^{-1} become stronger with respect to the GaAs-like LO mode, which is unable to see, as shown in Figure 5.13(b). As x is lowered further, the related GaAs-like LO can't be distinguished from the Raman spectrum in Figure 5.13(c) for sample #325 with a low Ga composition of x = 9.4%, and it possesses strong InAs-like LO at 236 cm^{-1} and AlAs-like LO at 270 cm^{-1}. Polarization measurements for all three samples confirmed that the Raman selection rule from the (100) surface of cubic crystals is obeyed.

5.8 RAMAN SCATTERING OF QUATERNARY InGaAlP

Another type of quaternary III-V compound InGaAlP materials, especially In$_{0.5}$(Ga$_{1-x}$Al$_x$)$_{0.5}$P which is lattice matched with GaAs, are important materials for visible red-green light-emitting diode and laser diode, solar cell, and other optoelectronic and electronic device applications. For example, many automobiles are equipped with back brick red lamps, which are made using InGaAlP materials. A set of six In$_{0.48}$(Ga$_{1-x}$Al$_x$)$_{0.52}$P/GaAs samples with the Al composition near 18% were prepared on GaAs by low-pressure metalorganic chemical vapor deposition (LP-MOCVD) [5.34], and another set of In$_{0.5}$(Ga$_{1-x}$Al$_x$)$_{0.5}$P thin films on GaAs substrates with a wide range of x up to ~80% were also grown by LP-MOCVD [5.35]. Both sets were studied using Raman scattering and related techniques by us [5.34, 5.35]. Also, Raman investigation on InGaAlP is performed similarly by other researchers [5.36, 5.37].

Figure 5.14 presents two typical Raman spectra of InGaAlP/GaAs, with different growth pressures, (a) sample S1 with a low growth pressure of 25 Torr and (b) S5 with 50 Torr (both at a H$_2$ flow of 85 slm). Both spectra show the allowed phonon modes of AlP–LO, GaP–LO, and InP-LO (the GaP-LO is sharpest), the forbidden InP–TO phonon, and GaAs-TO from the substrate.

To accurately obtain the peak energies, linewidths, FWHM, intensity, and background the Raman spectra, I(ω), were fitted with a broadened Lorentzian function,

$$I(\omega) = I_0 + \sum_{i=1}^{5} I_i \frac{\Gamma_i}{(\omega - \omega_i)^2 + \Gamma_i^2} \tag{5.7}$$

where I_0 is the background term (base line), ω_i and Γ_i are the peak position and FWHM, respectively, and i denotes InP–TO, InP–LO, GaP–LO, AlP–TO, and AlP–LO. A typical analyzing example for sample S5 is shown in Figure 5.15, where the open squares are the experimental data, the dotted lines are the individual components of Eq. (5.7), and the solid line is the total fit.

The analytical data from the Raman spectra in Figure 5.14 for this set of InGaAlP films are obtained, including the frequency and intensity of the InP-like TO mode, the frequencies of InP-like and AlP-like LO, the frequency, width, and intensity of GaP-like LO, the integrated intensity ratio between the allowed GaP–LO phonon, and the forbidden TO phonon, I(GaP–LO)/I(TO) [5.34]. In Figure 5.14(a), S1 with a low pressure of 25 Torr and a H$_2$ flow of 85 slm, the forbidden TO is stronger than all the allowed LOs in line intensity,

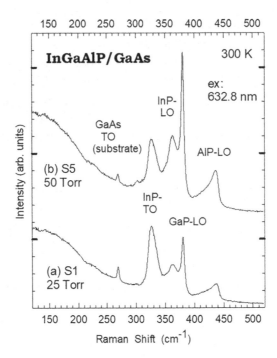

FIGURE 5.14   Raman spectra at 300 K of $In_{0.48}(Ga_{1-x}Al_x)_{0.52}P/GaAs$ (x ~ 0.18) samples: (a) S1 and (b) S-5.

Source: From [5.34], figure 2, with reproduction permission from AIP.

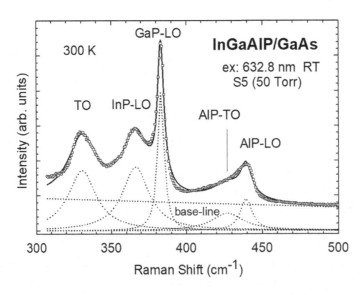

FIGURE 5.15   RT Raman spectrum and Lorentzian fit for $In_{0.48}(Ga_{1-x}Al_x)_{0.52}P/GaAs$ (x ~ 0.18) sample S-5. The open squares are the experimental data, and the solid line is the total fit result with the five Lorentzian components represented by the dashed lines.

Source: From [5.34], figure 3, with reproduction permission from AIP.

indicating a poor quality of this sample. Increasing the pressure to 50 Torr and keeping the same H_2 flow of 85 slm, for sample S5, all the allowed LOs in line intensity are increased with respect to the forbidden TO, indicating a good quality of this sample. Raman spectrum in Figure 5.14(b) with the GaP–LO mode much stronger than the TO phonon in intensity and the highest I(GaP–LO)/I(TO) ratio of 2.31, indicating the best growth conditions among these six samples [5.34].

In [5.35], we reported the growth and investigation for In0.5(Ga$_{1-x}$Alx)0.5P thin films on GaAs substrates with a wide range of x up to ~80% by LP-MOCVD. A variety of nuclear science and optical techniques, including Rutherford backscattering spectrometry, Raman scattering, PL, photoreflectance, and FTIR were employed to analyze these materials. A series of Raman spectra from MOCVD-grown InGaAlP/GaAs with the different Al compositions of x(Al) = 0.02, 0.24, 0.70, and 0.80, at two rotation degree directions, respectively, are exhibited. Raman phonon modes of GaAs LO and TO from GaAs substrate, quaternary InGaAlP alloy modes of AlP-like, GaP-like, InP-like LOs, and TOs, are studied correspondingly.

Our works provide evidence that the Raman spectral features are sensitive to the sample growth parameter variations. The line shape analysis of line width and integrated intensity ratio leads to information about the order of the sample crystalline quality and the optimum growth conditions that have been obtained from our comprehensive analyses.

REFERENCES

[5.1] Z. C. Feng, S. Perkowitz, T. S. Rao, and J. B. Webb, "Raman characterization of InSb/GaAs grown by metalorganic magnetron sputtering", MRS Online Proc. Libr. 160, 739–744 (1989). https://doi.org/10.1557/PROC-160-739.

[5.2] Z. C. Feng, S. Perkowitz, T. S. Rao, and J. B. Webb, "Resonance Raman scattering from epitaxial InSb thin films", J. Appl. Phys. **68**, 5363–5365 (1990). https://doi.org/10.1063/1.347033.

[5.3] Z. C. Feng, C. Beckham, P. Schumaker, I. Ferguson, R. A. Stall, N. Schumaker, M. Povloski, and A.Whitley, "Optical characterization and mapping of four inch InSb epitaxial thin films grown on GaAs by turbo disk metalorganic chemical vapor deposition", MRS Online Proc. Libr. **450**, 61–66 (1996). https://doi.org/10.1557/PROC-450-61.

[5.4] S. R. Das, C. Akatay, A. Mohammad, M. R. Khan, K. Maeda, R. S. Deacon, K. Ishibashi, Y. P. Chen, T. D. Sands, M. A. Alam, and D. B. Janes, "Electrodeposition of InSb branched nanowires: Controlled growth with structurally tailored properties", J. Appl. Phys. **116**, 083506 (2014). https://dx.doi.org/10.1063/1.4893704.

[5.5] A. P. Singh, K. Roccapriore, Z. Algarni, R. Salloom, T. D. Golden, and U. Philipose, "Structure and electronic properties of InSb nanowires grown in flexible polycarbonate membranes", Nanomaterials **9**, 1260 (2019). https://dx.doi.org/10.3390/nano9091260.

[5.6] I. Kh. Mammadov, D. H. Arasl, R. N. Rahimov, and A. A. Khalilova, "Raman scattering in the InSb – MnSb eutectic composite", Semiconductors **54**, 412–416 (2020). https://dx.doi.org/10.1134/S1063782620040089.

[5.7] Y. Qian, K. Xu, L. Cheng, C. Li, and X. Wang, "Rapid, facile synthesis of InSb twinning superlattice nanowires with a high-frequency photoconductivity response", RSC Adv. **11**, 19426 (2021). https://dx.doi.org/10.1039/d1ra01903a.

[5.8] Z. C. Feng, F. C. Hou, J. B. Webb, Z. X. Shen, E. Rusli, I. T. Ferguson, and W. Lu, "Optical investigation of GaSb thin films grown on GaAs by metalorganic magnetron sputtering", Thin Solid Films **516**, 5493–5497 (2008). https://doi.org/10.1016/j.tsf.2007.07.049.

[5.9] C. C. Ahia, N. Tile, E. L. Meyer, and J. R. Botha, "Effect of InSb deposition time on low-temperature photoluminescence and room temperature Raman of MOVPE grown InSb/GaSb nanostructures", Physica E **123**, 114197 (2020). https://doi.org/10.1016/j.physe.2020.114197.

[5.10] M. V. Kazimov, "Synthesis and structural analysis of InSb-CrSb, InSb-Sb, GaSb-CrSb eutectic composites", J. Optoelectron. Biomed. Mater. **12**, 67–72 (2020).

[5.11] D. N. Talwar, H.-H. Lin and Z. C. Feng, "Phonon characteristics of heavily Si-doped InAs grown by gas source molecular beam epitaxy", J. Raman Spectrosc. **50**, 1735–1743 (2019). https://doi.org/10.1002/jrs.5703.

[5.12] A. G. Lind, T. P. Martin, Jr., V. C. Sorg, E. L. Kennon, V. Q. Truong, H. L. Aldridge, C. Hatem, M. O. Thompson, and K. S. Jones, "Activation of Si implants into InAs characterized by Raman scattering", J. Appl. Phys. **119**, 095705 (2016). http://dx.doi.org/10.1063/1.4942880.

[5.13] V. K. Gupta, A. A. Ingale, S. Pal, R. Aggarwal, and V. Sathe, "Spatially resolved Raman spectroscopy study of uniform and tapered InAs micro-nano wires: Correlation of strain and polytypism", J. Raman Spectrosc. **48**, 855–866 (2017). https://doi.org/10.1002/jrs.5138.

[5.14] J. H. Park and C.-H. Chung, "Raman spectroscopic characterizations of self-catalyzed InP/InAs/InP one dimensional nanostructures on InP(111)B substrate using a simple substrate-tilting method", Nanoscale Res. Lett. **14**, 355 (2019). https://doi.org/10.1186/s11671-019-3193-6.

[5.15] Z. C. Feng, S. Perkowitz, F. W. Adams, F. Junga, D. Kinell, and R. Whitney, "Optical characterization of AlGaAs films grown on GaAs by molecular beam epitaxy", Mater. Res. Soc. Extend. Abs. **21**, 241–244 (1990).

[5.16] Z. C. Feng, S. Perkowitz, D. K. Kinnel, R. L. Whitney, and D. N. Talwar, "Compositional dependence of optical phonon frequencies in $Al_xGa_{1-x}As$", Phys. Rev. B **47**, 13466–13470 (1993). https://doi.org/10.1103/physrevb.47.13466.

[5.17] Y. T. Hou, Z. C. Feng, M. F. Li, and S. J. Chua, "Characterization of MBE grown $Ga_{1-x}Al_xAs$ alloy films by Raman scattering", Surf. Interface Anal. **28**, 163–165 (1999). https://doi.org/10.1002/(SICI)1096-9918(199908)28:1<163::AID-SIA598>3.0.CO;2-I.

[5.18] P. V. Seredin, A. V. Glotov, E. P. Domashevskaya, I. N. Arsentyev, D. A. Vinokurov, and I. S. Tarasov, "Raman investigation of low temperature AlGaAs/GaAs(1 0 0) heterostructures", Physica B **405**, 2694–2696 (2010). http://dx.doi.org/10.1016/j.physb.2010.03.049.

[5.19] J. Díaz-Reyes, M. Galván-Arellano, and R. Peña-Sierra, "Raman scattering and electrical characterization of AlGaAs/GaAs rectangular and triangular barriers grown by MOCVD", Superficies y Vacío **23**, 13–17 (2010). https://api.semanticscholar.org/CorpusID:54198891.

[5.20] R. S. Castillo-Ojeda, J. Díaz-Reyes, M. G. Arellano, M. de la Cruz Peralta-Clara, and J. S. Veloz-Rendón, "Growth and characterization of $Al_xGa_{1-x}As$ obtained by metallic-arsenic-based-MOCVD", Mater. Res. **20**, 1166–1173 (2017). http://dx.doi.org/10.1590/1980-5373-MR-2016-0512.

[5.21] Z. C. Feng, A. A. Allerman, P. A. Barnes, and S. Perkowitz, "Raman scattering of $In_xGa_{1-x}As$/InP grown by uniform radial flow epitaxy", Appl. Phys. Lett. **60**, 1848–1850 (1992). http://dx.doi.org/10.1063/1.107187.

[5.22] M.-S. Park, M. Razaei, K. Barnhart, C. L. Tan, and H. Mohseni, "Surface passivation and aging of InGaAs/InP heterojunction phototransistors", J. Appl. Phys. **121**, 233105 (2017). http://dx.doi.org/10.1063/1.4986633.

[5.23] R. Fujisawa, T. Iwamura, P. Leproux, V. Couderc, H. Okada, and H. Kano, "Ultrabroadband multiplex coherent anti-stokes Raman scattering (CARS) microspectroscopy using a CCD camera with an InGaAs image intensifier", Chem. Lett. **47**, 704–707 (2018). https://doi.org/10.1246/cl.180017.

[5.24] W. He, B. Li, and S. Yang, "High-frequency Raman analysis in biological tissues using dual-wavelength excitation Raman spectroscopy", Appl. Spectrosc. **74**, 241–244 (2020). https://doi.org/10.1177/0003702819881762.

[5.25] Z. C. Feng, S. Perkowitz, R. Rousina, and J. B. Webb, "Raman and infrared spectroscopies of $In_{1-x}Ga_xSb$ thin films on GaAs grown by metalorganic magnetron sputtering", Can. J. Phys. **69**, 386–389 (1991). https://doi.org/10.1139/p91-064.

[5.26] M. Macler, Z. C. Feng, S. Perkowitz, R. Rousina, and J. B. Webb, "Far infrared analysis of In_{1-x} Ga_xSb thin films on GaAs grown by metalorganic magnetron sputtering", Phys. Rev. B **46**, 6902–6906 (1992). https://doi.org/10.1103/PhysRevB.46.6902.

[5.27] I. Roh, S. H. Kim, D.-M. Geum, W. Lu, Y. H. Song, J. A. del Alamo, and J. D. Song, "High hole mobility in strained In0.25Ga0.75Sb quantum well with high quality Al0.95Ga0.05Sb buffer layer", Appl. Phys. Lett. **113**, 093501 (2018). https://doi.org/10.1063/1.5043509.

[5.28] J. E. F. Mena, R. C. Ojeda, and J. D. Reyes, "InSb czochralski growth single crystals for InGaSb substrates", Mater. Res. Soc. Symp. Proc. **1616**, 234 (2014). https://doi.org/10.1557/opl.2014.234.

[5.29] N. Nishimoto and J. Fujihara, "Characterization of a flexible InGaSb/PI thin film grown by RF magnetron sputtering and aqueous stability improvement via surface coating", Phys. Stat. Sol. A **2018**, 1800860 (2018). https://doi.org/10.1002/pssa.201800860.

[5.30] M. A. Ochoa, J. E. Maslar, and H. S. Bennett, "GaSb band-structure models for electron density determinations from Raman measurements", J. Appl. Phys. **135**, 0140357 (2023). https://doi.org/10.1063/5.0140357.

[5.31] Z. C. Feng, S. J. Chua, A. Raman, and K. P. J. William, "Growth and investigation of quaternary III-III-III-V InGaAlAs layers on InP by molecular beam epitaxy", J. Chin. Inst. Electr. Eng. **2**, 69–73 (1995). https://doi.org/10.1109/EDMS.1994.771204.

[5.32] K.-I. Min, H. Rho, J. D. Song, W. J. Choi, and Y. T. Lee, "Raman studies of InGaAlAs digital alloys", J. Korean Phys. Soc. **59** (4), 2801–2805 (2011). https://doi.org/10.3938/jkps.59.2801.

[5.33] M. Dang, H. Deng, S. Huo, R. R. Juluri, A. M. Sanchez, A. J. Seeds, H. Liu, and M. Tang, "The growth of low-threading dislocation-density GaAs buffer layers on Si substrates", J. Phys. D Appl. Phys. **56**, 405108 (2023). https://doi.org/10.1088/1361-6463/ace36d.

[5.34] Z. C. Feng, E. Armour, I. Ferguson, R. A. Stall, L. Malikova, T. Holden, J. Z. Wan, F. H. Pollak, and M. Pavlosky, "Non-destructive assessment of InGaAlP films grown on GaAs by low pressure metalorganic chemical vapor deposition", J. Appl. Phys. **85**, 3824–3831 (1999). https://doi.org/10.1063/1.369752.

[5.35] L. Li, C.-J. Hong-Liao, Y. Z. Huang, C. Chen, S. Yao, Z. R. Qiu, H. H. Lin, I. T. Ferguson, and Z. C. Feng, "Nuclear science and optical studies of InAlGaP materials grown on GaAs by metalorganic chemical vapor deposition", SPIE **8484**, 8484OY (2012). https://doi.org/10.1117/12.928862.

[5.36] M. Kondow, S. Minagawa, and S. Satoh, "Raman scattering from AlGaInP", *Appl. Phys. Lett.* **51**, 2001–2003 (1987). https://doi.org/10.1063/1.98273.

[5.37] X. Guo, A. Mughal, D. Dunphy, G. Stone, D. Miller, and S. Huh, "Ambient temperature – corrected mechanical stress mapping of gallium nitride and aluminum indium gallium phosphide films by Raman scattering spectroscopy for characterization of light-emitting diodes", Phys. Status Sol. A **2020**, 1900776 (2020). https://doi.org/10.1002/pssa.201900776.

CHAPTER 6

II-VI Semiconductors

6.1 RESONANT RAMAN SCATTERING OF CdTe

We have conducted the Raman and resonant Raman scattering investigation for CdTe thin films grown on GaAs by pulsed laser evaporation and epitaxy (PLEE) [6.1]. Raman scattering studies on CdTe were performed in [6.2–6.4]. Cadmium telluride (CdTe) is an important II-VI multifunctional semiconductor, often used as the light absorption layer in photovoltaic solar cells or the core layer in γ-ray detection. CdTe is an ideal photovoltaic cell material with a suitable direct band gap (1.45 eV) and a high absorption coefficient (absorb 99% of the sunlight with 1 μm thickness). It can be easily grown on different foreign substrates [2.3].

Figure 6.1 presents Raman and resonant Raman spectra of PLEE-grown (001) CdTe film (1.62 μm thick) on (001) GaAs at 80 K, excited with eight Ar$^+$ laser lines. The eight Raman spectra, excited by laser lines with energies of 2.4090–2.7067 eV (514.5–457.9 nm), exhibit a variety of sharp lines at multiples of the longitudinal optical (LO) phonon frequency, and broad bands [6.1].

The sharp Raman lines can be explained by considering three cases: no resonance (NR), incoming resonance (IR), and outgoing resonance (OR), defined relative to the $E_o + \Delta_o$ gap by Eqs. (6.1–6.3), respectively:

$$E_{in} < E_o + \Delta_o \, (NR), \tag{6.1}$$

$$E_{in} \sim E_o + \Delta_o \, (IR), \tag{6.2}$$

$$E_{out} = E_{in} - m\hbar\omega_{LO} \sim E_o + \Delta_o \, (OR). \tag{6.3}$$

The OR is the case that the Raman scattering efficiency has a maximum when the energy E_{out} of the scattered photon matches some energy gap E_g in the band structure. If an incoming photon of energy E_{in} is scattered m times by LO phonons, the final outgoing photon is emitted with energy $E_{out} = E_{in} - m\hbar\omega_{LO}$, giving Eq. (6.3) as the condition for OR with $Eg = E_o + \Delta_o$. In CdTe at 80 K, $E_o = 1.58$ eV, $E_o + \Delta_o = 2.535$ eV and $\hbar\omega_{LO} = 21.2$ meV [6.1]. Figure 6.1(a)–(c) shows the non-resonant case $E_{in} < E_o + \Delta_o$ (NR), with the first-order LO

DOI: 10.1201/9781032644912-6

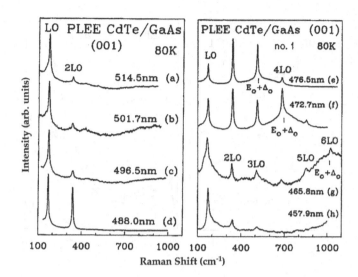

FIGURE 6.1   Raman and resonant Raman spectra of PLEE-grown (001) CdTe film (1.62 μm thick) on (001) GaAs at 80 K, excited with eight Ar⁺ laser lines.

Source: From [6.1], figure 1, with reproduction permission of APS.

Raman line much stronger than the second-order 2LO line, and the 3LO line hardly visible. Figure 6.1(d), with E_{in} of 2.541 eV (488 nm), approaching $E_o + \Delta_o = 2.535$ eV, that is, the onset of IR, having the 2LO line comparable to the LO line. As E_{in} increases further, other strong, sharp peaks appear at multiple LOs as shown in Figure 6.1(e)–(h) that can be explained by the OR relation, Eq. (6.3). In Figure 1(e), for example, with E_{in} =2.6013 eV, Eq. (6.3) is nearly exactly satisfied for m = 3. Therefore, the 3LO line resonates with the $E_o + \Delta_o$ gap and is extremely strong, only slightly less than that of the 2LO line. The 4LO also appears.

Figure 6.2 shows outgoing multi-phonon resonant Raman spectra of another (001) CdTe film (2.4 μm thick) on (001) GaAs measured at 80 K, excited by four Ar⁺ laser lines. The CdTe LO phonon overtones are superimposed on the strong $E_o+\Delta_o$ PL band in four cases, and an E_o PL band is inserted within (e).

We have also employed Raman scattering in the investigation for CdTe thin films grown on CdS/SnO₂/glass for solar cell application [6.5]. To investigate the correlation between the cell efficiency and material properties, Raman spectroscopy measurements were performed. Higher efficiency CdTe cells, grown by close-spaced sublimation and by metal-organic chemical vapor deposition (MOCVD) under Te-rich growth ambient, showed a strong Te band near 124 cm⁻¹ and CdTe TO mode near 142 cm⁻¹. A strong CdTe transverse optical (TO) mode indicates better CdTe quality and preferred [111] grain orientation. In the case of low-efficiency cells with Cd-rich CdTe, we saw a broader TO peak along with the emergence of LO mode near 168 cm⁻¹, suggesting lower film quality or multiple grain orientations.

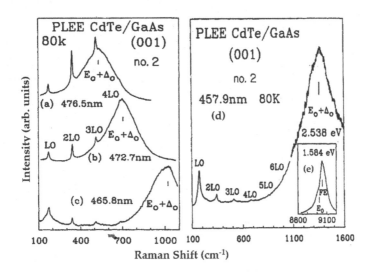

FIGURE 6.2   Outgoing multi-phonon resonant Raman spectra of PLEE-grown (001) CdTe film (2.4 μm thick) on (001) GaAs measured at 80 K, excited by four Ar⁺ laser lines.

Source: From [6.1], figures 2 and 3, with reproduction permission of APS.

6.2 RAMAN SCATTERING OF CdS

We have performed a Raman scattering investigation on CdS [6.6]. Raman scattering was used to study CdS materials in the literature [6.7–6.10]. Cadmium sulfide (CdS) is an n-type II–VI semiconductor with a direct energy band gap of 2.42 eV. CdS has been studied for decades owing to its applications in solar cells, transistors, and other devices. Particularly, thin CdS films have attracted much attention for their desirable characteristics of high light transparency in the visible part of the solar spectrum, enhanced photoconductivity, and easy ohmic contact formation with other materials. Thin CdS films can be fabricated by different deposition techniques [6.8]. CdS thin film as the window layer in CdTe/CdS-based II–VI solar cells have been attracting much research attention worldwide. We have studied CdTe/CdS/SnO$_2$/glass solar cells with CdS layers prepared by the solution method, and CdTe films subsequently grown by MOCVD [6.6].

Figure 6.3 shows Raman spectra of eight CdS films grown on SnO$_2$/glass. All the Raman spectra for eight CdS films show the Raman LO phonon band at ~305 cm⁻¹, which is characteristic of single- or poly-crystal CdS [6.6]. These Raman data for eight CdS films in Figure 6.3 are grouped into two, (a), (c), (e), and (g) for #11, #13, #15, and #17 without CdCl$_2$ treatments, while (b), (d), (f), and (h) for #12, 14, 16 and #18 with CdCl$_2$ treatments, respectively. Curve fittings were performed for all spectra, to get precise data values of the Raman peak frequencies and widths for comparisons. After CdCl$_2$ treatments, the characteristic CdS Raman LO mode became very narrow, indicating an improvement in the crystalline perfection of the CdS films.

Narrow Raman band and X-ray diffraction (XRD) peaks both indicate high crystalline perfection, fewer structural defects, and large crystalline zones. Sample #12 shows the narrowest Raman band and the narrowest peaks among the XRD bands (shown in

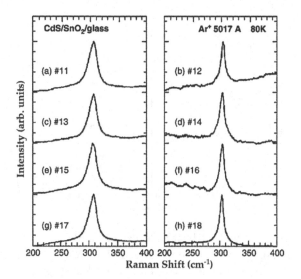

FIGURE 6.3   Raman spectra of CdS films grown on SnO_2/glass.

Source: From [6.6], figure 2, with reproduction permission of Elsevier.

[6.6]). Therefore, sample #12 can be identified as the best-quality sample among these eight CdS films. Sample #18 has the second narrowest Raman band and three first or second narrowest XRD peaks among the four. It is concluded that the optimum conditions for solution growth of CdS thin film on SnO_2-coated glass are S-rich CdS growth, followed by $CdCl_2$ treatment and annealing at 450°C in N_2. CdS growth with Cd-acetate plus $CdCl_2$ treatment and annealing leads to the second-best CdS film quality. Moreover, the observation of asymmetric Raman LO mode on its lower wave number side for mainly un-treated samples is related to the presence of defects and disorders in the crystal structure. The structure disorders allow contributions from the zone edge phonon to Raman spectra due to the breakdown of the crystalline symmetry. After $CdCl_2$ treatment and annealing, these disorder-related features have been greatly depressed or removed, indicating the improvement of CdS film crystalline perfection due to these treatments, which would be advantageous for application as the window layer in the solar cells [6.6].

6.3 RAMAN SCATTERING OF ZnTe

We have performed a Raman scattering investigation on ZnO and multiple phone overtones [6.11]. Other Raman studies on ZnTe are seen in [6.12–6.14]. ZnTe, a zincblende II-VI semiconductor, possesses a direct band gap in the green spectral region [6.11]. Zinc telluride and other II-VI semiconductors and their alloys are promising materials in the fabrication of light-emitting devices in the visible and near-ultraviolet optical range [6.13].

Figure 6.4 shows Raman spectra of bulk Bridgman-grown ZnTe at 80 K, excited at 514.5 nm and 100 mW in [I], with different polarization arrangements, and in [II], with different excitations of 501.7, 488.0, and 465.8 nm, respectively. Some ranges of spectra are magnified relative to the intensity of the free exciton band respectively, to see the multiple

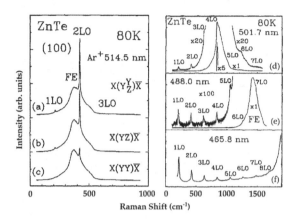

FIGURE 6.4 Raman spectra of bulk ZnTe at 80 K, excited at 514.5 nm and 100 mW in left graph, with the polarization arrangements of (a) unpolarized, (b) X(YZ)X̲, and (c) X(YY)X̲; and in right graph, excited at (d) 501.7, (e) 488.0, and (f) 465.8 nm. Partial spectra are magnified by factors of 5 and 20 in (d), and 100 in (e), relative to the intensity of the free exciton band respectively, to see the multiple phonons clearly.

Source: From [6.11], figures 1 and 2, with reproduction permission of Elsevier.

phonons clearly. We have examined resonance Raman scattering from bulk Bridgman-grown ZnTe, under excitations in 2.409–2.707 eV. Up to the eighth harmonic of the LO phonon peak appeared, due to the OR of LO phonons with the fundamental band gap. We compare the observed dependence of the intensity of the harmonics on order number m to a simple formula derived from the basic theory of Raman scattering, which includes a resonant term and a decreasing m^{-4} terms. Polarization data supports resonance Raman scattering as the origin of the multi-phonon lines.

As shown in Figure 6.4 left graph displaying unpolarized and polarized Raman spectra excited at 514.5 nm, the broad band centered at 2,363 eV is the free excite (FE) luminescence, only a few meV below the fundamental conduction valence E_0 band gap of ZnTe. Figure 6.4 (d) and (e) show clearly that the FE band is unpolarized. However, the 1LO, 2LO, and 3LO phonon lines, superimposed on the FE band, are partly polarized. These LO overtones do not come from the FE luminescence, but from resonance Raman scattering, not hot luminescence [6.11]. Figure 6.4 (a)–(b) shows also that the 2LO line is much stronger than the 1LO and 3LO lines. This is due to OR, like Section 6.1, where an incoming photon of energy E_{in} is scattered m times by LO phonons, giving a final E_{out} energy,

$$E_{out} = E_{in} - m\hbar\omega_{LO} \sim E_o\,(OR). \tag{6.4}$$

When E_{out} approximates a gap energy E_0 resonance ensues, enhancing the mLO lines. For the case of Figure 6.4 (a)–(c), $E_{in} = 2.409$ eV (514.5 nm), $E_g = E_0 = 2.363$ eV, and $m\hbar\omega_{LO} = 26.0$ meV, Eg. (6.4) predicts m = 1.8, near the observed value of 2. For the cases of Figure 6.4 (d) and (e), where higher-order peaks (although partly obscured by the free exciton peak) with m = 4 and 7 are excited at higher values of E_{in}. The mLO lines are all much weaker

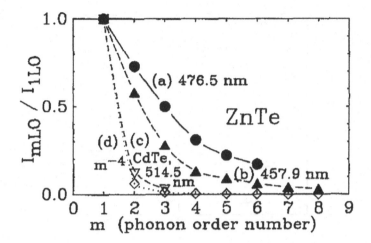

FIGURE 6.5   Integrated intensity I_{mLO} of mLO phonon lines far from the excitation versus order number m for excitations at (a) 476.5 and (b) 457.9 nm for ZnTe, and (c) 514.5 nm for CdTe with data cited from [6.1]. The line intensities are normalized relative to I_{1LO}. A plot of m^{-4} is shown in (d) for reference.

Source: From [6.11], figure 3, with reproduction permission of Elsevier.

than the strong FE transition but are still clearly visible with our sensitive Raman instrument. In Figure 6.4 (f), E_{in} is very high. The resonant peak lies far beyond the exciting line and outside our spectral range. Nevertheless, a series of overtones is visible, up to m = 8, with a striking monotonic decrease in intensity.

We have explored some expressions describing the resonant behavior and the monotonic decay far from resonance. The intensity of a mLO line is determined by the product of a term which resonates at $m\omega_{LO}$ and a term which decays rapidly as m^{-4}, and the I_{mLO}/I_{1LO} values are very close to m^{-4} [6.11]. Figure 6.5 shows the relationship of the I_{mLO}/I_{1LO} versus order number m for excitations at (a) 476.5 and (b) 457.9 nm for ZnTe, and (c) 514.5 nm for CdTe with data cited from [6.1]. The line intensities are normalized relative to I_{1LO}. A plot of m^{-4} is shown in (d) for reference. Our observation of LO overtones up to m = 8 in ZnTe, at many different excitation frequencies, provides good data to understand OR in semiconductors.

6.4 RAMAN SCATTERING OF CdMnTe

We have conducted a Raman scattering investigation on ternary CdMnTe [6.15, 6.16]. Similar studies of Raman scattering on CdMnTe are reported by others [6.17–6.19]. Dilute magnetic semiconductors (DMSs), also known as semimagnetic semiconductors, have attracted worldwide interest and research activity. Attention has focused on II-VI ternary alloys, where some of the group-II atoms are replaced at random by the magnetic transition-metal atom Mn. $Cd_{1-x}Mn_xTe$ (CMT) is the most extensively studied of these. Bulk CMT takes on the zincblende structure with x up to 0.77, and the wurtzite structure with larger x. The unfilled 3d shell of Mn produces localized magnetic moments which interact with the conduction and valence band electrons. The resulting novel magnetic and

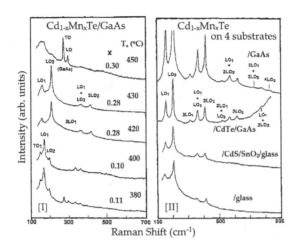

FIGURE 6.6   Raman spectra at 80 K and under 514.5 nm excitation from [I] $Cd_{1-x}Mn_xTe/GaAs$ samples grown at substrate temperatures T_s = 380–450°C, with x values between 0.11 and 0.30, and [II] $Cd_{1-x}Mn_xTe$ films grown at four different substrates of GaAs, CdTe/GaAs, CdS/SnO$_2$/glass, and glass, respectively.

Source: From [6.15], figures 2 and 3, with reproduction permission of AIP.

magneto-optical properties make CMT a promising material for electronic and photonic applications [6.16]. Also, CdMnTe materials are promising candidates for use in radiation detection as they exhibit several desirable characteristics, including an appropriate segregation coefficient (k ~ 1) for Mn, a low probability of crystal twins and Te precipitates, a wide bandgap, a high free-carrier mobility lifetime product, high resistivity, and excellent electronic transport properties [6.17].

Figure 6.6 shows Raman spectra with the 514.5 nm excitation and at 80 K, from [I] $Cd_{1-x}Mn_xTe/GaAs$ samples grown by MOCVD at substrate temperatures T_s = 380°C, 400°C, 420°C, 430°C, and 450°C, with x values of 0.11, 0.10, 0.28, 0.28, and 0.30, respectively; and [II] $Cd_{1-x}Mn_xTe$ films grown on four different substrates of GaAs, CdTe/GaAs, CdS/Sn02/glass, and glass [6.15]. The CdTe-like (LO_1) and MnTe-like (LO_2) modes as well as combinations and overtones are labeled and shown in spectra. Our results show that the MOCVD growth of CMT films on different substrates, including inexpensive glass, can be a successful technology.

Figure 6.7 exhibits Raman spectra at 80 K for [I] six PLEE-grown $Cd_{1-x}Mn_xTe$ (x = 0.058, 0.089, 0.44, 0.49, 0.54 and 0.70) films on GaAs under excitations of (a) 647.1, (b) 488.0, (c)–(e) 501.7, and (f) 457.9 nm, and [II] $Cd_{0.46}Mq_{0.54}Te/GaAs$, under six different exciting lines of 514.5, 501.7, 496.5, 488.0, 476.5 and 465.8 nm from an Ar$^+$ laser, respectively. LO phonon modes of CdTe-like LO_1 and MnTe-like LO_2 are labeled, together with their combinations of $LO_1 + LO_2$ and overtones.

We observed high-order (> 4) LO phonon overtones from $Cd_{1-x}Mn_xTe$ films with x(Mn) between 0.4 and 0.7 [6.16]. As shown in Figure 6.7 [II] for a $Cd_{0.46}Mq_{0.54}Te$ film, under the excitation of 501.7, 496.5, and 488.0 nm, the third- to sixth-order multiple LO phonon

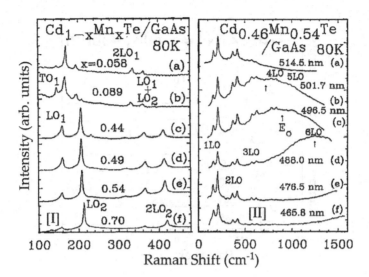

FIGURE 6.7 [I]-left: First- and second-order optical phonon Raman spectra at 80 K from six $Cd_{1-x}Mn_xTe$ films (x = 0.058–0.70) on GaAs substrate, with excitations at (a) 647.1 nm, (b) 488.0 nm, (c)-(e) 501.7 nm, and (f) 457.9 nm; [II]-right: Raman spectra at 80 K from $Cd_{0.46}Mq_{0.54}Te/GaAs$ under six different exciting lines from an Ar^+ laser, respectively.

Source: From [6.16], figures 1 and 3, with reproduction permission of AIP.

combinations are superposed upon the broad PL band, E_o, which can be explained by the concept of OR Raman scattering, discussed in Section 6.1 and [6.1]. The appearance of high-order phonon overtones and combinations characterizes the good crystalline structure of studied epitaxial $Cd_{1-x}Mn_xTe$ samples with a wide range of Mn-compositions.

6.5 RAMAN SCATTERING OF CdZnTe

We have performed Raman scattering studies of CdZnTe [6.20–6.23]. Raman research on CdZnTe was reported in [6.24–6.26]. The ternary $Cd_{1-x}Zn_xTe$ alloys are useful for a wide range of applications including terrestrial photovoltaics, nuclear medicine, digital radiography, electro-optical modulators, sensors, and detectors for hard x-rays and soft γ-rays. The cadmium-zinc-tellurides (CZTs) have emerged as one of the most important materials due to their intriguing characteristics and superiority over the traditional group IV elemental semiconductors. While the CZT-based sensors are imperative to identify radio-isotopes from the emitted γ-ray energies with better than 1% resolution at 662 keV [6.22].

Figure 6.8 shows Raman scattering spectra measured at 80 K and with the excitation of 488 nm from an Ar^+ laser, of bulk $Cd_{1-x}Zn_xTe$ with full range composition x(Zn) of (a) 1, (b) 0.5, (c) 0.4, (d) 0.3, (e) 0.2, (f) 0.1, (g) 0.03, (h) 0.01, and (i) 0.005. The laser power is 100 mW, focused on the sample surface with a spot size of about 0.2 mm.

In the frequency ω below 210 cm⁻¹, there are the first-order LO phonon modes of CdTe-like LO_1 and ZnTe-like LO_2 as well as their variations with x(Zn). In the frequency range of 320 cm⁻¹ < ω < 420 cm⁻¹, the second-order phonon modes of $2LO_1$, $2LO_2$, and LO_1+LO_2

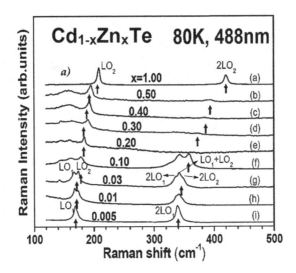

FIGURE 6.8   Raman scattering spectra of Bridgman-grown $Cd_{1-x}Zn_xTe$ ternary alloys of different compositions, x, measured at 80 K and under the 488 nm excitation.

Source: From [6.23], figure 2, with reproduction permission of Elsevier.

are presented. It is interesting to find that at lowest x = 0.005, the CdTe-like mode LO_1 and the ZnTe-like model LO_2 are mixed, with the LO_2 mode appearing as a weak shoulder at the high-energy side of the LO_1. As x increases to 0.01, the LO_1 and LO_2 modes are clearly resolved in Figure 6.8(h). For x = 0.03 at Figure 6.8(g), and x = 0.10 at Figure 6.8(f), these modes are well separated, indicating the two-mode behavior clearly. Till x(Zn) = 1, that is, pure ZnTe, only ZnTe-like LO_2 and $2LO_2$ appear. Because of the (100) surface of our samples, only LO phonons are allowed in the Raman measurement geometry. Low-temperature Raman scattering study has not only offered an excellent testimony to the lattice phonons of the perfect materials. Our comprehensive experimental and theoretical studies have apprehended the vibrational and related properties of Bridgman-grown $Cd_{1-x}Zn_xTe$ alloys [6.20–6.23].

6.6 RAMAN SCATTERING OF CdSeTe

We have conducted a combined Raman, infrared, photoluminescence, and theoretical investigation on the II-VI-VI ternary CdSeTe [6.27, 6.28]. Raman scattering and related studies on CdSeTe were also reported [6.29, 6.30]. Ternary alloyed $CdSe_{1-x}Te_x$ (abbreviated as CdSeTe) exhibits a so-called bowing effect; that is, the plot of the band gap versus the composition is concave upward. Therefore, the band gap of CdSeTe can be smaller than that of constituent binary parents CdSe and CdTe. Due to the bowing effect, the light absorption range of CdSeTe can be shifted to the near-infrared region. This property can be exploited for the construction of solar cells and near-infrared luminescent probers for in vivo molecular imaging and biomarker detection and so on [6.29].

Figure 6.9 presents Raman spectra of four Bridgman-grown bulk $CdSe_xTe_{1-x}$ crystals, measured at 80 K and excited by the 501.7 nm line from an Ar^+ laser, with values of x

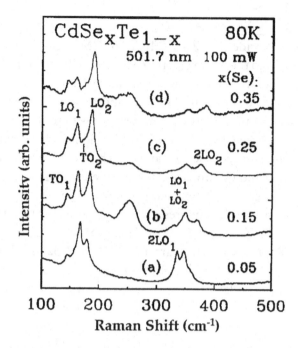

FIGURE 6.9   Raman spectra of Bridgman-grown bulk $CdSe_xTe_{1-x}$ at 80 K, excited by 501.7 nm, with values of x being (a) 0.05, (b) 0.15, (c) 0.25, and (d) 0.35.

Source: **From [6.27], figure 1, with reproduction permission of Elsevier.**

being (a) 0.05, (b) 0.15, (c) 0.25, and (d) 0.35. Raman spectra in the frequency range of 140 and 200 cm^{-1} have revealed CdTe-like and CdSe-like TO and LO phonons, labeled TO_1, TO_2, LO_1, and LO_2, respectively. Even in a sample with the lowest value of x = 0.05 in Figure 6.9(a), the CdTe-like LO_1, TO_1, and CdSe-like LO_2, TO_2 (very weak) modes are distinguishable; the LO_2 mode appears as a shoulder at the high-energy side of the LO_1. As x increases to 0.15, the LO_1 and LO_2 modes become well separated, in Figure 6.9(b).

Low-temperature (80 K) Raman spectra have revealed the classic CdTe-like (TO_1, LO_1) and CdSe-like (TO_2, LO_2) pairs of optical phonons, and the composition-dependent peak shifts and intensity variations. As x increases between 0.05 and 0.35, the LO_2 mode peak frequencies exhibited shifts toward the higher-energy side, while those of the LO_1 phonon frequencies have unveiled slight redshifts. At the same time, the intensity of the LO_2 line increases and LO_1 decreases. The dispersion relationship of four first-order Raman peak positions at 80 K versus x for $CdSe_xTe_{1-x}$ (x = 0–1) is displayed graphically in Figure 2 of [6.27].

Further, the observed vibrational features between 300 and 400 cm^{-1} in Figure 6.9 are due to the second-order LO phonons, that is, the combinations of LO_1 and LO_2, $2LO_1$, LO_1 + LO_2, and $2LO_2$. In compound semiconductors, the strength of higher-order phonons is generally sensitive to the degree of crystalline perfection. As x increases, the intensities of the second-order LO phonon feature relative to the first-order peaks decrease, indicating an increased disorder in the experimental samples. Also, the broad feature near 250 cm^{-1} in Figure 6.9 is perhaps due to defects [6.27, 6.28].

6.7 RAMAN SCATTERING OF ZnMnTe

The zinc–manganese–telluride (i.e., $Zn_{1-x}Mn_xTe$ or ZMT) can be prepared by alloying manganese (Mn^{2+}) ions into the nonmagnetic ZnTe lattice. By varying the Mn composition x in ZnTe helped tune the lattice parameters and energy band gaps to prepare ZMT nanostructures, for example, quantum wells (QWs), quantum dots, and superlattices (SLs). The presence of magnetic ions in II-VI lattices generally leads to some unusual properties due to spin–spin exchange between d-electrons of the Mn^{2+} ions and s-electrons of the conduction band and p-electrons of the valence bands. Such interactions between the localized magnetic moments and band electrons are responsible for typical magneto-optical traits having extremely large Zeeman splitting and Faraday rotation [6.31].

Our collaborative team has performed research on "Optical and Structural Characteristics of Bridgman-Grown Cubic $Zn_{1-x}Mn_xTe$ Alloys" [6.31]. Studies on $Zn_{1-x}Mn_xTe$ and related materials, including Raman scattering, are reported in [6.32–6.34]. Figure 6.10 shows Raman spectra, measured in the backscattering geometry at room temperature and under the excitation of 780 nm of Bridgman-grown cubic $Zn_{1-x}Mn_xTe$ crystals with different Mn composition of x(Mn) = 0.003, 0.06, 0.12, and 0.15, respectively [6.31].

The MnTe-like modes are marked with TO_1 at ~195 cm^{-1} and LO_1 at ~211 cm^{-1}, while the ZnTe-like optical phonon modes of TO_2 at 177 cm^{-1} and LO_2 at 207 cm^{-1} are identified. The observed spectrum of a sample with the lowest Mn composition x = 0.003 is found quite like that of the binary ZnTe. On the top spectrum of x = 0.15, besides observing long-wavelength transverse optical (TO_2) and longitudinal optical (LO_2) modes, six (labeled 1–6) major phonon features displayed at the low- and high-frequency sides are attributed to the two-mode combinations.

These assignments were made by using the rigid ion model phonons fitted to the inelastic neutron scattering data, incorporating appropriate selection rules. As x is changed from

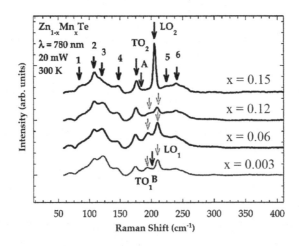

FIGURE 6.10   Room-temperature Raman spectra of $Zn_{1-x}Mn_xTe$ samples for different Mn composition of x(Mn) = 0.003, 0.06, 0.12, and 0.15, measured in the backscattering geometry with λ = 780 nm.

Source: From [6.31], figure 2, with reproduction permission of Elsevier.

$0.06 \rightarrow 0.12 \rightarrow 0.15$, we noticed (a) the low- and high-frequency Raman features of the two-mode combinations – exhibiting similar behavior as seen in a sample of $x = 0.003$ with the exception of perceiving small frequency shifts and variation in their intensities, (b) the observation of MnTe-like TO_1 and LO_1 modes shown, and (c) the weak disordered induced features A and B near ~180 and 205 cm^{-1} are attributed to the impurity vibrational modes of MnTe:Zn and ZnTe:Mn, respectively. Experimental and theoretical investigations are reported to comprehend the structural and lattice-dynamical properties of high-quality Bridgman-grown cubic $Zn_{1-x}Mn_xTe$ ($x \leq 0.15$) crystals. Our detailed Raman scattering and far-infrared reflectivity measurements ascertained a typical "intermediate-phonon-mode" behavior for the alloys [6.31].

6.8 RAMAN SCATTERING OF ZnMnSe

$Zn_{1-x}Mn_xSe$ is a wide-bandgap ternary material, belonging to the II-Mn-VI group of semi-magnetic or DMSs. It has an energy gap tunable between 2.7 and 3.4 eV, depending on the Mn composition and the applied magnetic field. Magnetic ions, Mn^{2+}, are randomly substituted with the Zn^{2+} in the cation positions, leading to interesting magneto-optical properties, such as a giant Zeeman splitting of the band edges. This has been utilized to obtain spin-dependent quantum confinement in tailored structures, promising for next-generation electro-optical and photonic devices [6.35].

Our collaborative team has conducted a combined Raman and infrared spectroscopic investigation on ZnMnSe materials [6.35]. Other similar studies, including Raman scattering, on ZnMnSe, were reported in the literature [6.36, 6.37]. Figure 6.11 presents Raman scattering spectra of $Zn_{1-x}Mn_xSe$ for different x values at 300 K. A series of epitaxial $Zn_{1-x}Mn_xSe$ on GaAs with Mn composition variation from 0% up to 78% were grown by the molecular beam epitaxy (MBE) technique.

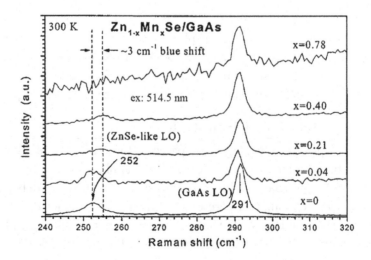

FIGURE 6.11 Raman scattering spectra of MBE-grown $Zn_{1-x}Mn_xSe$ on GaAs for different x values at 300 K and under the 514.5 nm excitation.

Source: From [6.35], figure 3, with reproduction permission of APS.

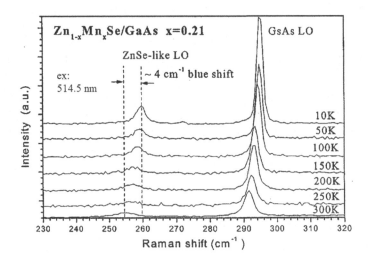

FIGURE 6.12   Temperature-dependent Raman spectra of a $Zn_{1-x}Mn_xSe/GaAs$ with x = 0.21 from 10 to 300 K.

Source: From [6.35], figure 4, with reproduction permission of APS.

The Raman peak at 291 cm^{-1} is the LO phonon from the GaAs substrate. The ZnSe Raman LO mode appears at 252 cm^{-1}, as the LO$_1$ mode at x = 0. With increasing x value from x = 0 to 0.52, the ZnSe-like LO$_1$ vibration mode has a 3 cm^{-1} blueshift, and the mode intensity weakens. For x = 0.78, the LO$_1$ mode is too weak to be recognized. The LO$_1$ mode also shows the temperature dependence that all $Zn_{1-x}Mn_xSe$ epilayers with different Mn concentrations have a blueshift of 3–4 cm^{-1} when temperature varies from 300 K down to 10 K.

Figure 6.12 shows temperature-dependent Raman spectra of a $Zn_{1-x}Mn_xSe/GaAs$ with x = 0.21 from 10 to 300 K and under the 514.5 nm excitation. Raman data provide experimental evidence on the intermediate-mode phonon behavior for $Zn_{1-x}Mn_xSe$ with x up to 0.78, and temperature dependence of a typical $Zn_{1-x}Mn_xSe/GaAs$ with x = 0.21 from 10 to 300 K. Furthermore, the relationships of optical mode (ZnSe-, MnSe-like TO and LO) frequency versus x full range from epitaxial $Zn_{1-x}Mn_xSe$ are plotted in Figure 5 of [6.35].

6.9 RAMAN SCATTERING OF HgTe-CdTe SL

The HgTe/CdTe SLs possess advantages compared to the $Hg_{1-x}Cd_xTe$ alloy of the same band gap, very suitable for far-infrared optoelectronics applications [6.38]. There is a small valence band offset V_0 and a large spin-orbit splitting in the HgTe-CdTe heterostructures [6.39]. We have performed Raman and resonant Raman scattering for the HgTe/CdTe SLs [6.38]. Studies on HgTe and HgTe/CdTe SLs, including Raman scattering and related techniques, were presented in [6.39–6.41]. Figure 6.13 shows Raman spectra of two HgTe/CdTe SLs measured at 10 K and excited under 476.4 nm.

For sample No. 2 with 300-Å CdTe cap, like the case of Figure 6.1(e), CdTe 1–4 LOs are observed in Figure 6.13 (b), due to the OR with $E_0 + \Delta_0$ of the CdTe cap. These CdTe multiple LO harmonics are strongly enhanced relative to the other peaks. Raman spectrum

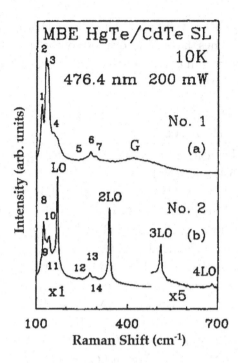

FIGURE 6.13   Raman spectra of two HgTe/CdTe superlattices at 10 K, excited at 476.4 nm. (a) SL sample No. 1, with 150 periods of [HgTe (64Å)]/[CdTe (60Å)] and with no CdTe cap. (b) SL sample No. 2, with 300-Å CdTe cap, 150 periods of [HgTe (80Å)]/[CdTe (40Å)], and 3.6-μm CdTe buffer, on CdTe substrate.

Source: From [6.38], figure 1, with reproduction permission of APS.

at Figure 6.13 (a) for SL No. 1 shows HgTe-related lines with virtually no CdTe lines. The HgTe TO and LO phonon peaks are prominent at 118 cm^{-1} (peak 1) and 138 cm^{-1} (peak 3), respectively. The HgTe-like LO line from $Hg_{0.15}Cd_{0.85}Te$ appears at 132 cm^{-1} (peak 2). The HgTe-like TO mode at 127 cm^{-1} is submerged in the low-energy tail of the pure HgTe and HgTe-like LO bands. Second-order phonon combinations appear over 240–340 cm^{-1}, where the peak 6 at 278 cm^{-1} is the HgTe 2LO mode. The peak 7 at 295 cm^{-1} is the combined CdTe LO + HgTe-like LO line. A broad feature (labeled G) extends from 350 to 550 cm^{-1}, with a peak near 430 cm^{-1}, like the case reported by P. J. Olego et al. [6.39], not due to photoluminescence, but unknown for the origin yet. Another unknown Raman band is seen at 155 cm^{-1}.

Figure 6.13(b) presents the Raman spectrum at 476.4 nm excitation of HgTe/CdTe SL No. 2, with the CdTe cap layer, which shows how the increased strength in the CdTe modes minimizes the HgTe peaks. Alloy HgTe-like modes like those in No. 1 still appear for SL No. 2. Peak 8 (126 cm^{-1}) and peak 9 (133 cm^{-1}) lie at the HgTe-like TO- and LO phonon frequencies, respectively, from $Hg_{0.15}Cd_{0.85}Te$. The CdTe-like LO mode from $Hg_{0.15}Cd_{0.85}Te$ appears at 165 cm^{-1} (peak 11), as a shoulder of the strong pure CdTe LO band. The CdTe-like TO mode, which should appear near 144 cm^{-1}, is submerged by the pure CdTe TO band (peak 10, 143 cm^{-1}). However, the HgTe-like modes are dwarfed by the CdTe LO and

2LO modes. The latter may be large simply because the CdTe cap layer absorbs much of the incoming light before it reaches the SL. The locations of the $Hg_{1-x}Cd_xTe$ phonon lines confirm the degree of Hg alloying in the nominal CdTe layers, showing the utility of Raman scattering for SL characterization.

6.10 RAMAN SCATTERING OF CdTe-CdMnTe MULTIPLE QUANTUM WELL

Our collaborative team has carried on a Raman scattering of CdTe-CdMnTe multiple quantum well (MQW) [6.42]. Research on Raman scattering of CdTe-CdMnTe MQWs [6.43] and CdTe-CdMgTe QWs [6.44] were also reported. Figure 6.14 exhibits the Raman spectra for a CdTe-$Cd_{0.90}Mn_{0.10}$Te MQW sample CCM-109, measured under excitation of (a) 501.7 and (b) 476.5 nm, respectively. LO_{1b}, TO_{1b}, LO_{2b}, and $2LO_{1b}+LO_{2b}$ are from $Cd_{0.90}Mn_{0.10}$Te barriers and cap, while LO_w, $2LO_w$, and $3LO_w$ are from CdTe wells.

By the pulsed laser evaporation and epitaxy (FLEE) technology, (001) CdTe-$Cd_{1-x}Mn_xTe$ (x = 0.10) MQW and SL structures were grown on (001) $Cd_{0.95}Zn_{0.05}$Te substrates. The sample CCM-109 consists of 22 pairs of CdTe-$Cd_{0.90}Mn_{0.10}$Te QWs, with a 1500 Å $Cd_{1-x}Mn_xTe$ buffer layer and capped with a 225 Å thick $Cd_{0.90}Mn_{0.10}$Te layer. For non-resonance excitation of 501.7 nm at Figure 6.14(a), CdTe-like LO phonon from $Cd_{0.90}Mn_{0.10}$Te barriers and cap LO_{1b} (at 166 cm^{-1}), MnTe-like phonon LO_{2b} (at 194 cm^{-1}), the overtone $2LO_{1b}$ and the combination $LO_{1b} + LO_{2b}$ dominate the spectrum. The CdTe-like TO phonon TO_{1b} (at 144 cm^{-1}) is also indicated. The CdTe LO phonon from the wells has emerged in the strong LO_{1b} band and appears relatively weak as a high wavenumber shoulder of LO_{1b}. An interface mode (IF) located below LO_{2b} is also exhibited weakly in the spectrum.

For resonance excitation of 476.5 nm at Figure 6.14(b), the CdTe phonons from CdTe wells, LO_w at 170 cm^{-1}, and its overtones of $2LO_w$ at 340 cm^{-1} and $3LO_w$ at 510 cm^{-1} are

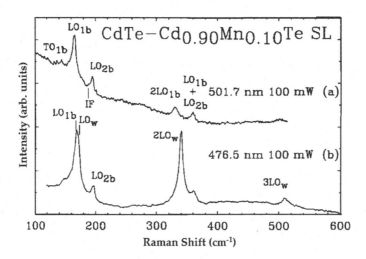

FIGURE 6.14   Raman spectra for a CdTe-$Cd_{0.90}Mn_{0.10}$Te superlattice (SL) measured under excitation of (a) 501.7 and (b) 476.5 nm.

Source: From [6.42], figure 3, with reproduction permission of Elsevier.

greatly excited. This is like the case of outgoing multiple LO phonon resonance Raman scattering from PLEE-grown CdTe films [6.1], discussed in Section 6.1. Raman scattering for resonant conditions shows LO phonons and their overtones up to the third-order which have originated in the CdTe wells. Also, CdTe-like LO_{1b}, LO_{2b}, and $LO_{1b} + LO_{2b}$ from the $Cd_{0.90}Mn_{0.10}Te$ barriers are displayed.

REFERENCES

[6.1] Z. C. Feng, S. Perkowitz, J. M. Wrobel, and J. J. Dubowski, "Outgoing multi-phonon resonant Raman scattering and luminescence near the $E_o+\Delta_o$ gap in epitaxial CdTe films", Phys. Rev. B **39**, 12997–13000 (1989). https://doi.org/10.1103/PhysRevB.39.12997.

[6.2] S. Sohal, M. Edirisooriya, T. Myers, and M. Holtz, "Investigation of cadmium telluride grown by molecular-beam epitaxy using micro-Raman spectroscopy below and above the laser damage threshold", J. Vac. Sci. Technol. B **36**, 052905 (2018). https://doi.org/10.1116/1.5048526.

[6.3] X. Ai, S. Yan, Y. Chen, S. Chen, Y. Jiang, X. Song, L. Ma, and X. Wu, "Microstructure and optical properties of CdTe thin films prepared by close spaced sublimation method at various growth temperatures", J. Lumin. **252**, 119372 (2022). https://doi.org/10.1016/j.jlumin.2022.119372.

[6.4] S. Surabhi, K. Anurag, and S. R. Kumar, "Structural, morphological, electrical and optical properties of Zn doped CdTe thin films", Chalcogenide Lett. **19**, 143–152 (2022). https://doi.org/10.15251/CL.2022.192.143.

[6.5] A. Rohatgi, H. C. Chou, N. M. Jokerst, E. W. Thomas, C. Ferekides, S. Kamra, Z. C. Feng, and K. M. Dugan, "Effects of CdTe growth conditions and techniques on the efficiency limiting defects and mechanisms in CdTe solar cells", AIP Conf. Proc. **353**, 368–375 (1996). https://doi.org/10.1063/1.49428.

[6.6] Z. C. Feng, C. C. Wei, A. T. S. Wee, A. Rohatgi, and W. Lu, "Effects of $CdCl_2$ treatment and annealing on CdS/SnO_2/glass heterostructures for solar cells", Thin Solid Films **518**, 7199–7203 (2010). https://doi.org/10.1016/j.tsf.2010.04.076.

[6.7] G. L. Song, S. Guo, X. X. Wang, Z. S. Li, B S Zou, H. M. Fan, and R. B. Liu, "Temperature dependent Raman and photoluminescence of an individual Sn-doped CdS branched nanostructure", New J. Phys. **17**, 063024 (2015). https://doi.org/10.1088/1367-2630/17/6/063024.

[6.8] M. F. Saleem, H. Zhang, Y. Deng, and D. Wang, "Resonant Raman scattering in nanocrystalline thin CdS film", J. Raman Spectrosc. **48**, 224–229 (2016). https://doi.org/10.1002/jrs.5002.

[6.9] R. Tan, D. F. Kelley, and A. M. Kelley, "Resonance hyper-Raman scattering from CdSe and CdS nanocrystals", J. Phys. Chem. C **123**, 16400–16405 (2019). https://doi.org/10.1021/acs.jpcc.9b04645.

[6.10] F. A. Pisu, P. C. Ricci, S. Porcu, C. M. Carbonaro, and D. Chiriu, "Degradation of CdS yellow and orange pigments: A preventive characterization of the process through pump – probe, reflectance, X-ray diffraction, and Raman spectroscopy", Materials **15**, 5533 (2022). https://doi.org/10.3390/ma15165533.

[6.11] Z. C. Feng, S. Perkowitz, and P. Becla, "Multiple phonon overtones in ZnTe", Solid State Commun. **78**, 1011–1014 (1991). https://doi.org/10.1016/0038-1098(91)90120-K.

[6.12] J. Camacho, A. Cantarero, I. Hernandez-Calderon, and L. Gonzalez, "Raman spectroscopy and photoluminescence of ZnTe thin films grown on GaAs", J. Appl. Phys. **92**, 6014–6018 (2002). http://dx.doi.org/10.1063/1.1516267.

[6.13] R. S. Castillo-Ojeda, J. Díaz-Reyes, M. Galván-Arellano, F. de Anda-Salazar, J. I. Contreras-Rascon, M. de la Cruz Peralta-Clara, and J. S. Veloz-Rendón, "Structural characterization of ZnTe grown by atomic-layer-deposition regime on GaAs and GaSb (100) oriented substrates", Mater. Res. **20**, 1179–1184 (2017). http://dx.doi.org/10.1590/1980-5373-MR-2016-0181.

[6.14] E. V. Borisov, A. A. Kalinichev, and I. E. Kolesnikov, "ZnTe crystal multimode cryogenic thermometry using Raman and luminescence spectroscopy", Materials **16**, 1311 (2023). https://doi.org/10.3390/ma16031311.

[6.15] Z. C. Feng, R. Sudharsanan, S. Perkowitz, A. Erbil, K. T. Pollard, and A. Rohatgi, "Raman scattering characterization of high-quality $Cd_{1-x}Mn_xTe$ films grown by metalorganic chemical vapor deposition", J. Appl. Phys. **64**, 6861–6863 (1988). https://doi.org/10.1063/1.341977.

[6.16] Z. C. Feng, S. Perkowitz, and J. J. Dubowski, "Raman scattering studies of $Cd_{1-x}Mn_xTe$ films on GaAs by pulsed laser evaporation and epitaxy", J. Appl. Phys. **69**, 7782–7787 (1991). https://doi.org/10.1063/1.348926.

[6.17] J. Lai, J. Wang, L. Wang, H. Ji, R. Xu, J. Zhang, J. Huang, Y. Shen, J. Min, L. Wang, and Y. Xia, "Characterization of CdMnTe films deposited from polycrystalline powder source using closed-space sublimation method", J. Vac. Sci. Technol. A **33**, 05E125 (2015). http://dx.doi.org/10.1116/1.4927820.

[6.18] P. Yu, T. Shao, Z. Ma, P. Gao, B. Jing, W. Liu, C. Liu, Y. Chen, Y. Liu, Z. Fang, and L. Luan, "Influence of hydrogen treatment on electrical properties of detector-grade CdMnTe:In crystals", IEEE Trans. Nucl. Sci. **68**, 458–462 (2021). https://doi.org/10.1109/TNS.2021.3067726.

[6.19] P. Yu, T. Shao, W. Liu, P. Gao, B. Jiang, S. Zhao, Z. Han, X. Gu, and J. Zheng "Preparation and characterization of pure phase CdMnTe nanopowders by a hydrothermal route", RSC Adv. **12**, 19006–19015 (2022). https://doi.org/10.1039/d2ra02020c.

[6.20] S. Perkowitz, L. S. Kim, Z. C. Feng, and P. Becla, "Optical phonons in $Cd_{1-x}Zn_xTe$", Phys. Rev. B **42**, 1455–1457 (1990). https://doi.org/10.1103/physrevb.42.1455.

[6.21] D. N. Talwar, Z. C. Feng, and P. Becla, "Impurity-induced phonon disordering in $Cd_{1-x}Zn_xTe$ ternary alloys", Phys. Rev. B **48**, 17064–17069 (1993). https://doi.org/10.1103/PhysRevB.48.17064.

[6.22] D. Talwar, Z. C. Feng, J.-F. Lee, and P. Becla, "Extended x-ray absorption fine structure and micro-Raman spectra of Bridgman grown $Cd_{1-x}Zn_xTe$ ternary alloys", Mater. Res. Exp. **1**, 015018 (2014). http://doi.org/10.1088/2053-1591/1/1/015018.

[6.23] D. N. Talwar, P. Becla, H.-H. Lin, and Z. C. Feng, "Assessment of intrinsic and doped defects by optical spectroscopy in Bridgman grown $Cd_{1-x}Zn_xTe$", Mater. Sci. Eng. B **269**, 115160 (2021). https://doi.org/10.1016/j.mseb.2021.115160.

[6.24] Y. V. Znamenshchykov, V. V. Kosyak, A. S. Opanasyuk, V. O. Dord, P. M. Fochuk, and A. Medvids, "Raman characterisation of $Cd_{1-x}Zn_xTe$ thick polycrystalline films obtained by the close-spaced sublimation", Acta Phys. Pol. A **132**, 1430–1435 (2017). https://doi.org/10.12693/APhysPolA.132.1430.

[6.25] G. Kartopu, Q. Fan, O. Oklobia, and S. J. C. Irvine, "Combinatorial study of the structural, optical, and electrical properties of low temperature deposited $Cd_{1-x}Zn_xTe$ $(0 \leq x \leq 1)$ thin films by MOCVD", Appl. Surf. Sci. **540**, 148452 (2021). https://doi.org/10.1016/j.apsusc.2020.148452.

[6.26] T. Alhaddad, M. B. Shoker, O. Pagès, A. V. Postnikov, V. J. B. Torres, A. Polian, Y. Le Godec, J.-P. Itié, L. Broch, M. B. Bouzouraâ A. E. Naciri, S. Diliberto, S. Michel, P. Franchetti, A. Marasek, and K. Strzałkowski, "Raman study of $Cd_{1-x}Zn_xTe$ phonons and phonon-polaritons – experiment and *ab initio* calculations", J. Appl. Phys. **133**, 065701 (2023). https://doi.org/10.1063/5.0134454.

[6.27] Z. C. Feng, P. Becla, L. S. Kim, S. Perkowitz, Y. P. Feng, H. C. Poon, K. P. Williams, and G. D. Pitt, "Raman, infrared, photoluminescence and theoretical studies of the II-VI-VI ternary CdSeTe", J. Cryst. Growth **138**, 239–243 (1994). https://doi.org/10.1016/0022-0248(94)90814-1.

[6.28] D. N. Talwar, Z. C. Feng, J.-F. Lee, and P. Becla, "Structural and dynamical properties of Bridgeman grown $CdSe_xTe_{1-x}$ $(0 < x \leq 0.35)$ ternary alloys", Phys. Rev. B. **87**, 165208 (2013). https://doi.org/10.1103/PhysRevB.87.165208.

[6.29] L. X. Hung, P. T. Nga, N. N. Dat, and N. T. T. Hien, "Temperature dependence of Raman and photoluminescence spectra of ternary alloyed $CdSe_{0.3}Te_{0.7}$ quantum dots", J. Electron. Mater. **49**, 2568–2577 (2020). https://doi.org/10.1007/s11664-020-07961-x.

[6.30] A. Ciris, B. M. Bas, Y. Atasoy, A. Karaca, T. Kucukomeroglu, M. Tomakin, and E. Bacaksiz, "Deposition of CdSeTe alloys using CdTe-CdSe mixed powder source material in a close-space sublimation process", J. Mater. Sci. Mater. Electron. **32**, 9685–9693 (2021). https://doi.org/10.1007/s10854-021-05630-1.

[6.31] D. N. Talwar, P. Becla, H.-H. Lin, and Z. C. Feng, "Optical and structural characteristics of Bridgman grown cubic $Zn_{1-x}Mn_xTe$ alloys", Mater. Chem. Phys. **220**, 460–468 (2018). https://doi.org/10.1016/j.matchemphys.2018.07.042.

[6.32] V. S. Vinogradov, T. N. Zavaritskaya, G. Karczewski, I. V. Kucherenko, N. N. Melnik, and W. Zaleszczyk, "Raman scattering and hot luminescence spectra of $Zn_{1-x}Mn_xTe$ quantum wires", Phys. Solid State **52**, 1757–1762 (2010). https://doi.org/10.1134/S1063783410080299.

[6.33] V. F. Agekyan, E. V. Borisov, A. Y. Serov, N. G. Filosofov, and G. Karczewski, "Optical properties of ZnMnTe/ZnMgTe quantum_well nanostructures", Phys. Solid State **57**, 1831–1836 (2010). https://doi.org/10.1134/S1063783415090024.

[6.34] J. Dong, Y. Xu, L.-L. Ji, B. Xiao, B.-B. Zhang, L. Guo, C. Zhang, C. Teichert, and W. Jie, "Enhanced terahertz response of diluted magnetic semiconductor $Zn_{1-x}Mn_xTe$ crystals", Opt. Mater. Exp. **8**, 157–164 (2018). https://doi.org/10.1364/OME.8.000157.

[6.35] T. R. Yang, C. C. Lu, W. C. Chou, Z. C. Feng, and S. J. Chua, "Infrared and Raman spectroscopic study of ZnMnSe materials grown by molecular beam epitaxy", Phys. Rev. B **60**, 16058–16064 (1999). https://doi.org/10.1103/PhysRevB.60.16058.

[6.36] C.-T. Tsai, D.-S. Chuu, and J.-Y. Leou Chou, "fabrication and physical properties of radio frequency sputtered ZnMnSe thin films", Jpn. J. Appl. Phys. **36**, 4427–4430 (1997). https://doi.org/10.1143/JJAP.36.4427.

[6.37] P. J. Klar, P. J. Boyce, D. Wolverson, J. J. Davies, W. Heimbrodt, N. Hoffmann, and J. Griesche, "Spin-flip Raman scattering studies of $ZnSe/Zn_{1-x}Mn_xSe$ multiple quantum well structures", J. Cryst. Growth **159**, 1061–1065 (1996). https://doi.org/10.1016/0022-0248(95)00694-X.

[6.38] Z. C. Feng, S. Perkowitz, and O. K. Wu, "Raman and resonant Raman scattering for the HgTe/CdTe superlattice", Phys. Rev. B **41**, 6057–6060 (1990). https://doi.org/10.1103/PhysRevB.41.6057.

[6.39] D. J. Olego, J. P. Faurie, and P. M. Raccah, "Optical investigation of hole and electron subbands in Hg Te-CdTe superlattices", Phys. Rev. Lett. **55**, 328 (1985). https://doi.org/10.1103/PhysRevLett.55.328.

[6.40] M. Lv, R. Wang, L. Wei, G. Yu, T. Lin, N. Dai, J. Chu, and D. J. Lockwood, "Strained HgTe plates grown on SrTiO3 investigated by micro-Raman mapping", J. Appl. Phys. **120**, 115304 (2016). https://doi.org/10.1063/1.4962852.

[6.41] A. Laref, M. Alsagri, Z. A. Alahmed, and S. Laref, "First-principles analysis for the modulation of energy band gap and optical characteristics in HgTe/CdTe superlattices", RSC Adv. **9**, 16390 (2019). https://doi.org/10.1039/c8ra10101a.

[6.42] J. J. Dubowski, A. P. Roth. E. Deleporte, G. Peter, Z. C. Feng, and S. Perkowitz, "Optical properties of $CdTe-Cd_{0.90}Mn_{0.10}Te$ quantum well structures grown by pulsed laser evaporation and epitaxy", J. Cryst. Growth **117**, 862–866 (1992). https://doi.org/10.1016/0022-0248(92)90873-H.

[6.43] Y. G. Kusrayev, A. V. Koudinov, D. Wolverson, and J. Kossut, "Anisotropy of spin-flip Raman scattering in CdTe/CdMnTe quantum wells", Phys. Stat. Sol. B **229**, 741–744 (2002). https://doi.org/10.1002/1521-3951(200201)229:2 < 741::AID-PSSB741 > 3.0.CO;2-N.

[6.44] A. V. Koudinov, E. V. Borisov, A. A. Shimko, Y. E. Kitaev, C. Trallero-Giner, T. Wojtowicz, G. Karczewski, and S. V. Goupalov, "Ultranarrow lines in Raman spectra of quantum wells due to effective acoustic phonon selection by in-plane wave vector", Phys. Rev. B **105**, L121301 (2022). https://doi.org/10.1103/PhysRevB.105.L121301.

CHAPTER 7

ZnO-Based Semiconductors

7.1 RAMAN SCATTERING OF ZnO CRYSTAL

We have performed a series of Raman scattering studies on ZnO and alloys [7.1–7.11]. Raman scattering is an important and widely used technique in hot and frontier research of ZnO-based materials [7.12–7.20]. Zinc oxide (ZnO) has been recognized as the third class of promising wide-gap semiconductor together with SiC and GaN. It is an "old" semiconductor which has been compelling research attention for a long time because of its applications in scientific and industry areas such as piezoelectric transducers, optical waveguides, acousto-optic media, conductive gas sensors, transparent conductive electrodes, and varistors. ZnO, crystallizing in the wurtzite structure, is a direct band gap semiconductor with a room-temperature band gap of 3.37 eV, an exciton binding energy of 60 meV, and other good properties. ZnO normally forms in the hexagonal (wurtzite) crystal structure with lattice constants of a = 3.250 Å and c = 5.207 Å [7.8].

ZnO can be grown at relatively low growth temperatures below 500°C. The band gap of ZnO can be tuned via divalent substitution on the cation site to produce heterostructures. For example, Cd substitution leads to a reduction in the band gap to ~3.0 eV. ZnO with MgO is possible to tune the E_g from 3.37 eV (ZnO band gap) to 7.8 eV (MgO band gap). Substituting Mg on the Zn site in epitaxial films can increase the band gap to approximately 4.0 eV, while still maintaining the wurtzite structure. Therefore, ZnO and related materials as well as quantum/nanostructures have now received increasing attention and are recognized as promising candidates for efficient UV/blue light-emitting diodes, sensors, photodetectors, and laser diodes [7.8].

For the Raman scattering studies of ZnO, in the review chapter "Brief History Review of Research/Development and Basic/Interdisciplinary Characterization on ZnO", in my edited *Handbook of Zinc Oxides and Related Materials: Volume 1: Materials*, chapter 1, I have outlined the basic and important Raman characteristics: ZnO with Wurtzite structure and has C6v symmetry with six Raman-active phonon modes, namely, A1(longitudinal optical (LO)), A1 transverse optical (TO), E1(LO), E1(TO), E2(low), and E2(high), at frequencies of 574, 380, 583, 407, 101, and 438 cm⁻¹, respectively [7.8]. Two additional B modes are silence modes. The bulk ZnO crystal materials grown by modified

DOI: 10.1201/9781032644912-7

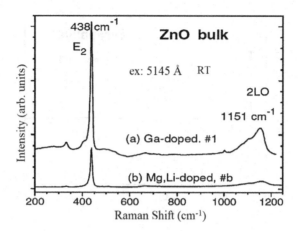

FIGURE 7.1 Raman spectra of two bulk ZnO, measured at room temperature and excited under 5145 Å.

Source: From [7.1], figure 1, with reproduction permission from Springer.

melt growth technique were studied by us via Raman and other characterization techniques [7.1, 7.2, 7.8].

Figure 7.1 presents Raman spectra of two bulk ZnO, measured at room temperature (RT) and under 5145 Å from an Ar$^+$ laser. Both exhibit a strong peak at 438 cm^{-1}, due to the ZnO E2 vibration, characteristic of the wurtzite structure, with high crystallinity. A broad peak appears at ~1151 cm^{-1} for the Ga-doped ZnO, and relatively weaker for the Mg, Li-doped bulk ZnO, which corresponds to the 2LO phonon mode. It is related to the coupling between the phonons and free carriers in ZnO [7.1]. These can be used to characterize and determine the free-carrier concentration and doping level in ZnO, like the case of SiC and GaN, discussed in Chapters 2 and 3.

7.2 RAMAN SCATTERING OF ZnO EPI-FILMS

Figure 7.2 shows the micro-Raman scattering spectra for three ZnO films, under the excitation of 488 nm. A sharp peak at 437 cm^{-1} is observed for all ZnO epi-films, which is the wurtzite ZnO characteristic E_2 mode, but shifted slightly due to the strain in the ZnO grown on sapphire substrate. The intensity and full width at half maximum (FWHM) of the E_2 mode were employed to represent the quality of ZnO materials. Higher intensity, narrow, and symmetric features indicate our metal-organic chemical vapor deposition (MOCVD)-grown ZnO films on sapphire substrate with high crystalline quality [7.3].

Figure 7.3 exhibits resonant Raman spectra of three MOCVD-grown ZnO under the excitation of 325 nm from a He-Cd laser. The ultraviolet (UV) 325 nm-excited resonant Raman scattering with a strong A_1 symmetry LO phonon mode, 1LO at 556 cm^{-1}, and its second-order 2LO mode was observed for all samples. The spectral rising in the high-frequency side in Figure 7.3 is from the ZnO edge mission photoluminescence (PL) band.

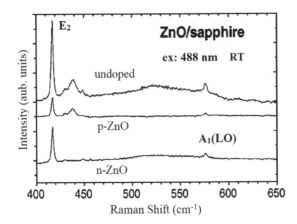

FIGURE 7.2 Raman spectra of three MOCVD-grown ZnO under the excitation of 488 nm from an Ar⁺ laser.

Source: From [7.3], figure 7, with reproduction permission from Elsevier.

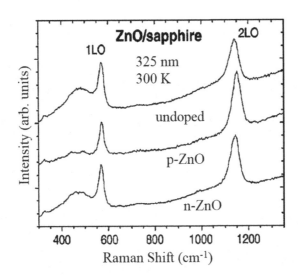

FIGURE 7.3 UV (325 nm)-excited Raman spectra of three MOCVD-grown ZnO films grown on sapphire.

Source: From [7.3], figure 8, with reproduction permission from Elsevier.

The A$_1$(LO) and 2LO phonon modes were enhanced due to the resonance of the UV 325 nm excitation (beyond the ZnO band gap) with the fundamental ZnO PL band. All results have shown the high crystalline quality of MOCVD-grown ZnO films, indicated by the narrow XRD, PL, and Raman line widths, strong PL signals, sharp optical transition edge, and smooth surface. A high p-type carrier concentration of up to 10^{19} cm^{-3} has been achieved besides the good n-type doping in ZnO. New classes of p–n junction devices based on ZnO including blue and UV LEDs, lasers, and detectors are achieved [7.3].

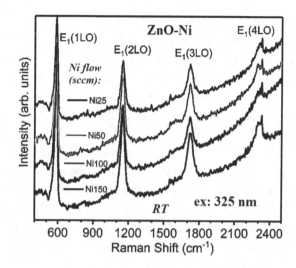

FIGURE 7.4   Resonant Raman spectra, at room temperature and with the excitation of 325 nm, for four MOCVD-grown ZnO doped with Ni under different Ni flows of 25, 50, 100, and 150 sccm, respectively.

Source: **From [7.11], figure 8, with reproduction permission from AIP.**

Figure 7.4 shows the 325 nm laser-excited Raman spectra of the Zinc nickel oxide (ZnO-Ni) thin films with an average thickness of 125 nm and four different Ni-flow rates (25, 50, 100, and 150 sccm, respectively), grown on sapphire c-plane (0001) substrates by MOCVD. The multiplications of the LO phonon mode $E_1(LO)$, labeled $E_1(1LO)$, $E_1(2LO)$, $E_1(3LO)$, and $E_1(4LO)$, at 584 cm^{-1}, 1168 cm^{-1}, 1752 cm^{-1}, and 2336 cm^{-1}, are seen, respectively. The risen backgrounds in Raman spectra start at 1500 cm^{-1}, which are broad PL band tails of ZnO with a peak near 380 nm, that is, around 4400 cm^{-1}. The four LO peaks are from resonant Raman scattering in ZnO-based materials, observed in the high-energy region of the ZnO PL peak. The 1LO peak at 584 cm^{-1} shows the asymmetry shape. Generally, the asymmetry of the Raman peak is due to a break in crystalline materials' translational symmetry at grain boundaries. This causes vibrational contributions from surfaces and interfaces and the formation of new phases that have spectral contributions. As also seen in HR-XRD analysis, Ni-related phases were formed in ZnO-Ni along with a reduction in crystal grains' size and an increase in grain boundaries, which could result in asymmetry in the Raman spectra. Detailed combining of Raman and PL phonon replica modes and coupling effects, as well as spectral asymmetry to the FWHM of Raman peaks related to Ni-dopings, are invested by our team [7.11].

7.3 UV RAMAN SCATTERING OF Cr- AND Cu-DOPED ZnO FILMS

Diluted magnetic semiconductors (DMSs), known as semiconductors doped with small concentrations of transition-metal ions, are receiving intense attention due to the possibility of utilizing both charge and spin degrees of freedom in the same materials. The realization of spintronic devices can be adopting the DMS nanowires or nanorods. In addition, ferromagnetic DMS nanowires and nanorods have been reported to have higher Curie temperature (Tc) and larger magnetic moments, compared to their bulk and film counterparts. In our collaborative research [7.9], $Zn_{0.92}Cu_{0.08}O$ nanorod arrays were synthesized by radio-frequency magnetron sputtering deposition at different substrate temperatures. Cu-doped ZnO nanorods showed ferromagnetism (FM) observed at 20°C. With the increase in substrate temperature, an improved crystallinity along with the increase in oxygen vacancies was observed in $Zn_{0.92}Cu_{0.08}O$ nanorods, which caused an enhancement of magnetic moment in $Zn_{0.92}Cu_{0.08}O$ nanorods. The experimental results showed that the interaction between substitutional Cu–Zn in a divalent charge and oxygen vacancies played an important role in the origin of ferromagnetism in $Zn_{0.92}Cu_{0.08}O$ nanorods.

Figure 7.5 presents UV (325 nm) excited combined PL and resonant Raman spectra of the $Zn_{0.92}Cr_{0.08}O$ nanorods deposited on p-type Si (111) substrates by radio-frequency magnetron sputtering with different substrate temperature (T_s) of 20°C, 150°C, and 600°C, respectively. The inset shows a Raman spectrum, with 6LO, 7LO, and 8LO of the $Zn_{0.92}Cr_{0.08}O$ grown at 20°C [7.9]. An ultraviolet (UV) peak near 383 nm (3.24 eV) is displayed in Figure 7.5. This UV emission band is related to a near band edge transition of ZnO, that is, the recombination of the free excitations. Their positions remain almost unchanged in different samples, but their intensities have a large difference. The intensity of the UV peak is evidently enhanced with the increase of the substrate temperatures. This indicates that the nonradiative centers relating to the Cu dopants (Cu acted as traps to the excited electrons) are decreased with increasing substrate temperature up to 600°C.

Resonant Raman scattering (RRS) is a good tool for studying the crystallinity and physical properties of semiconductors including nanostructure materials. In general, the number of LO phonons in semiconductors varies monotonically with the electron-LO phonon

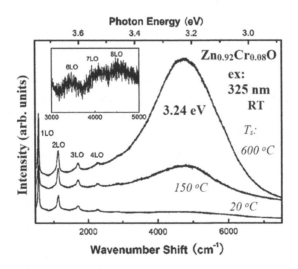

FIGURE 7.5 PL and resonant Raman spectra of the $Zn_{0.92}Cr_{0.08}O$ grown at 20°C, 150°C, and 600°C with excitation of 325 nm, respectively. The inset shows the partial Raman spectrum of the $Zn_{0.92}Cr_{0.08}O$ grown at 20°C, with 6LO, 7LO, and 8LO, respectively.

Source: From [7.9], figure 3, with reproduction permission from Elsevier.

coupling. It is seen from Figure 7.5 that, with the increase of the substrate temperatures, the intensities of the LO peaks increase, which is due to the increase of the oxygen vacancies. It shows that the Raman intensity of 1LO is larger than that of 2LO, which may be due to impurities or defective scatters. The I(2LO)/I(1LO) ratio increases monotonically with increasing the substrate temperatures, indicating the decrease of the disorder effect and other scatter effects on the RRS. When the sample grows at high substrate temperature, the grain size increases, and the crystallinity of the nanorods becomes better. Also, the electron-LO phonon coupling will be enhanced accordingly [7.9].

7.4 UV RAMAN SCATTERING OF NONPOLAR A-PLANE ZnO FILMS

As a direct wide-bandgap semiconductor, zinc oxide (ZnO) is a material with great potential for ultraviolet (UV) light-emitting diode (LED), modulators, and transistors. However, spontaneous and piezoelectric polarizations, which usually exist in conventionally grown c-oriented ZnO-based devices due to its wurtzite crystal structure, result in the strong built-in electrostatic fields and the overall reduction in radiative recombination efficiency, leading to poor performance. Therefore, the growth of ZnO-based devices along nonpolar directions (e.g., a- and m-directions) has attracted attention because they can effectively circumvent the adverse effects of undesirable spontaneous and piezoelectric polarizations. Ascribed to the intrinsic asymmetry of the wurtzite lattice structure, the a-plane ZnO films present an in-plane anisotropy of optical characteristics and have been used in potential applications such as UV modulators and novel polarization-sensitive devices [7.10 and references therein].

Our collaborative team has conducted comprehensive research on nonpolar a-plane ZnO epi-films grown on r-Sapphire substrate, a-GaN template, and a-$Al_{0.08}Ga_{0.92}N$ template, respectively [7.10]. The a-GaN and a-AlGaN substrates were prepared using MOCVD. Subsequently, three substrates were put into the reaction chamber of the pulsed laser deposition (PLD) system for ZnO deposition. Figure 7.6 shows the Raman spectra for three nonpolar a-ZnO layers. In Figure 7.6, two dashed lines at 379.0 and 439.0 cm^{-1} indicate the phonon frequencies of A_1(TO) and E_2(high), respectively, from strain-free ZnO. The corresponding frequency shifts of $\Delta\omega$ (A_1(TO)) and $\Delta\omega$ (E_2(high)) are labeled for each sample. The in-plane tensile strain ε_{yy} and in-plane compressive strain ε_{zz} were estimated and indicated in Figure 7.6. Between r-sapphire and a-ZnO, there exists a large lattice and thermal mismatch, while a-ZnO and a-GaN have similar lattice constants and a small difference between in-place linear thermal expansion coefficient. Therefore, the in-plane strains of a-ZnO grown on a-GaN and a-AlGaN are much smaller than that of a-ZnO grown on r-Sapphire. These are revealed by the frequency shifts of $\Delta\omega$ (E_2(high)) from curve (b) of a-ZnO/a-GaN and (c) of a-ZnO/a-AlGaN being much smaller than those of spectrum (a) a-ZnO/r-sapphire [7.10].

As the Al component x in a-$Al_xGa_{1-x}N$ template increases from 0 to 0.08, both in-plane tensile strain ε_{yy} and in-plane compressive strain increase. This should be attributed to the smaller lattice constant of AlN than GaN and ZnO. We can speculate that the ε_{yy} and ε_{zz} will continue to increase with the increase of x. This indicates that the in-plane strains of a-ZnO grown on a-$Al_xGa_{1-x}N$ can be reduced by increasing the Al component x, based on our Raman experimental outcome.

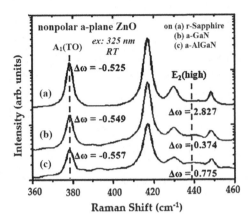

FIGURE 7.6 Raman scattering spectra for three nonpolar a-ZnO layers. $\Delta\omega$ (A_1(TO)) and $\Delta\omega$ (E_2(high)) are labeled for each sample.

Source: From [7.10], figure 4, with reproduction permission from The Optical Society.

7.5 SURFACE ENHANCED RAMAN SCATTERING OF ZnO NANOSTRUCTURES

Our collaborative team performed the surface-enhanced Raman scattering on ZnO nanostructures, formed by using the Langmuir–Blodgett technology [7.5]. Figure 7.7 presents Raman spectra of ZnO nanocrystals measured at RT and with different laser radiation wavelengths, that is, under resonance and non-resonance conditions.

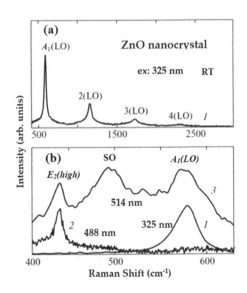

FIGURE 7.7 Raman spectra of the ZnO nanocrystal measured under (a) resonance and (b) non-resonance conditions and recorded at different excitation wavelengths of (1) 325 nm, (2) 488 nm, and (3) 514.5 nm, respectively.

Source: From [7.5], figure 4, with reproduction permission from Springer.

Figure 7.7(a) displays the RSS spectrum for the ZnO nanocrystal, excited under 325 nm and at RT. Multiple LO modes, including $A_1(LO)$, 2LO, 3LO, and 4LO are shown. The Raman spectrum, at RT and with the 514.5 nm excitation, as shown in Figure 7.7(b), exhibits a broad feature (approximately 50 cm^{-1} wide) near 495 cm^{-1}. It is located between the E_2(high) and $A_1(LO)$ modes and attributed to the surface optical (SO) phonons, which are not observed under the excitation of 488 and 325 nm. This SO mode has been reported in more ZnO nano-structural materials [7.5, 7.21–7.23].

7.6 RAMAN SCATTERING OF MgZnO EPI-FILMS

The band gap of ZnO can be widened without changing its crystal structure by incorporating Mg into the ZnO matrix. $Zn_{1-x}Mg_xO$ ternary alloy is considered the most suitable barrier layer for carrier confinement due to its similar lattice constant to that of ZnO, and its tunable wide band gap (in the range from 3.37 eV to about 4 eV, according to its composition x), and thus it is a good material for $ZnO/Zn_{1-x}Mg_xO$ superlattices and quantum wells. ZnO and its alloy with MgO have extensive application prospects in short-wavelength optoelectronic devices. Moreover, $Zn_{1-x}Mg_xO$ alloys can be successfully p-type doped, with phosphors (P), lithium (Li), antimony (Sb) monodoping, indium plus nitrogen (In–N), and boron plus phosphorus (B–P) co-doping methods. However, the fundamental studies on acceptor doping in $Zn_{1-x}Mg_xO$ thin films are crucial for the fabrication of good ZnO-based devices. [7.16].

Figure 7.8 shows the Raman scattering spectra of five $Mg_xZn_{1-x}O$ alloys with x(Mg) = 0–0.14 under 532 nm excitation and at RT. This series of $Mg_xZn_{1-x}O$ alloy (x = 0, 0.01, 0.06, 0.10 and 0.14) thin films were grown on a sapphire substrate by MOCVD. The upper insert figure amplifies the range of 560–595 cm^{-1}, for 1-LO mode. The ZnO-like first-order longitudinal optical (1-LO) phonon lines are located at almost the same frequency with no significant shift.

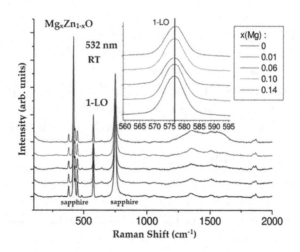

FIGURE 7.8 Raman scattering spectra of five $Mg_xZn_{1-x}O$ alloys with x(Mg) = 0–0.14 under 532 nm excitation. The insert figure amplifies the range of 560–595 cm^{-1}, for 1-LO mode.

Source: From [7.4], figure 1, with reproduction permission from AIP.

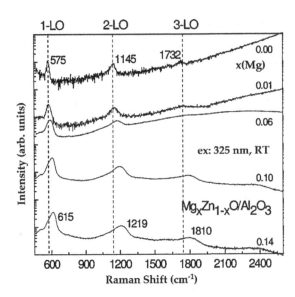

FIGURE 7.9 The UV 325 nm excited Raman spectra with $Mg_xZn_{1-x}O$ alloy in different x (Mg) content.

Source: From [7.4], figure 3, with reproduction permission from AIP.

Figure 7.9 displays the UV 325 nm excited Raman spectra of these five samples with $Mg_xZn_{1-x}O$ alloy in different x(Mg) content. Under the RRS, unlike the non-resonant case in Figure 7.8, the Raman LOpeak frequencies are changed with x(Mg). With the increase of x(Mg) from 0 to 0.14, the ZnO-like first-order (1LO) phonon line shows a significant shift from 575 to 615 cm^{-1}, the second-order LO (2LO) has a shift from 1145 to 1219 cm^{-1}, and the third-order (3LO) also shifts from 1732 to 1810 cm^{-1}. Also, with the increase of Mg content, the 1-LO is getting broadened, and so too are the 2-LO and 3-LO. It shows that the structure has been affected by the replacement of Zn^{2+} to Mg^{2+}. The quality of the ZnO wurtzite structure is getting worse as x (Mg) increases.

By way of the combined PL–Raman measurements excited by a 325 nm UV laser, PL peak shifts and multiple LO mode variations are observed, which indicate the variation and sensitivity of growth temperature on the MgZnO alloys. These relationships are practically useful in material structure characteristics such as when 1LO of Raman shift reaches 730 cm^{-1}, it means the ZnO hexagonal structure has turned into MgO cubic structure [7.4, 7.7].

7.7 RAMAN SCATTERING OF AlZnO EPI-FILMS

Al-doped ZnO (AZO) as a transparent conductive oxide with superior properties, can be a good alternative for tin-doped indium oxide (ITO) in optoelectronic applications such as LEDs, flat panel displays and solar cells [7.6, 7.24–7.26]. In the collaborative work, nanometer-scale AlZnO thin film materials epitaxial on sapphire substrates from PLD were investigated by synchrotron radiation (SR) X-ray absorption fine-structure spectroscopy, photoluminescence (PL) and Raman scattering [7.6, 7.24]. The AZO samples are prepared by PLD using ZnO:Al target at different temperatures ranging from 350°C to 650°C.

We have presented the RT Raman scattering measurement data of three AlZnO thin films grown on sapphire with different growth temperatures of 350°C, 550°C, and 650°C, respectively, as shown in Figure 7.3 of [7.6]. The dominant and prominent peaks of Al-doped ZnO films are E_2(low) mode at 101 cm^{-1} and E_2 (high) at 442 cm^{-1}, respectively. Several sharp peaks at 379, 419, 431, 449, 576, and 750 cm^{-1} are observed due to the sapphire substrate. For the undoped ZnO, the sharpest and strongest peak of E_2(low) and E_2(high) modes appeared at about 98 and 438 cm^{-1}, respectively, indicating the ZnO wurtzite crystal structure.

The strong and narrow E_2 (high) mode for all AlZnO films indicates good crystallinity, fewer structural defects, and large crystalline zones. To the wurtzite structure crystals, stress induced in the crystals affects the E_2 phonon frequency obviously, from which the information on stress can be extracted. This stress is from a mismatch between the thin film and the substrate, or the distortion. It is considered that doping is the factor that would cause the lattice distortion of the crystals, because of the difference between the atomic radius of different elements.

Al-doped ZnO grown at 350°C possesses weak/broad Raman signals indicating a poor crystalline film. The AlZnO film grown at a growth temperature of 550°C possesses the E_2(high) mode stronger and narrower than that of AlZnO grown at 650°C. The Raman experimental data reveals that the PLD-grown AlZnO film grown on sapphire could get better crystalline quality at 550°C than at 350°C and 650°C, which was further confirmed by the 266 nm excitation micro-PL and other measurements. The quality of crystallinity of AlZnO has been evaluated by means of Raman scattering.

REFERENCES

[7.1] S. Ganesan, Z. C. Feng, D. Mehta, M. H. Kane, I. Ferguson, J. Nause, B. Wagner, and C. Summers, "Optical properties of bulk and epitaxial ZnO", MRS Online Proc. Libr. **799**, 321–326 (2003). https://doi.org/10.1557/PROC-799-Z8.10.

[7.2] Z. C. Feng, J. W. Yu, J. B. Wang, R. Varatharajan, B. Nemeth, J. Nause, I. Ferguson, W. Lu, and W. E. Collins, "Optical characterization of ZnO materials grown by modified melt growth technique", Mater. Sci. Forum **527–529**, 1567–1570 (2006). https://doi.org/10.4028/www.scientific.net/MSF.527-529.1567.

[7.3] S. Sun, G. S. Tompa, C. Rice, X. W. Sun, Z. S. Lee, S. C. Lien, C. W. Huang, L. C. Cheng, and Z. C. Feng, "Metal organic chemical vapor deposition and investigation of ZnO thin films grown on sapphire", Thin Solid Films **516**, 5572–5577 (2008). https://doi.org/10.1016/j.tsf.2007.07.030.

[7.4] C. C. Wei, Y. L. Tu, Z. C. Feng, C. C. Wu, P. R. Lin, and D. S. Wuu, "Resonant Raman scattering of MgZnO thin films grown on sapphire by MOCVD", ICORS 2010, AIP Conf. Proc. **1267**, 1115–1116 (2010). https://doi.org/10.1063/1.3482327.

[7.5] A. G. Milekhin, N. A. Yeryukov, L. L. Sveshnikova, T. A. Duda, E. I. Zenkevich, S. S. Kosolobov, A. V. Latyshev, C. Himcinski, N. V. Surovtsev, S. V. Adichtchev, Z. C. Feng, C. C. Wu, D. S. Wuu, and D. R. T. Zahn, "Surface enhanced Raman scattering of light by ZnO nanostructures", J. Exp. Theor. Phys. **113**, 983–991 (2011). https://doi.org/10.1134/S1063776111140184.

[7.6] Y. R. Lan, S.-P. Liu, C. C. Wei, Y.-C. Fu, Y. R. Wu, J. M. Chen, M.-H. Wang, R.-H. Horng, D.-S. Wuu, and Z. C. Feng, "X-ray absorption fine-structure and optical studies of AlZnO nano-thin films grown on sapphire by pulsed laser deposition", SPIE **8104**, 81040X (2011). https://doi.org/10.1117/12.893087.

[7.7] C. C. Wu, D. S. Wuu, P. R. Lin, T. N. Chen, R. H. Horng, S. L. Ou, Y. L. Tu, C. C. Wei, and Z. C. Feng, "Characterization of $Mg_xZn_{1-x}O$ thin films grown on sapphire substrates by metalorganic chemical vapor deposition", Thin Solid Films **519**, 1966–1970 (2011). http://doi.org/10.1016/j.tsf.2010.10.036.

[7.8] Z. C. Feng, "Brief History Review of Research/Development and Basic/Interdisciplinary Characterization on ZnO", Chapter 1 in Handbook of Zinc Oxides and Related Materials: Volume 1. Materials, CRC Press, Taylor & Francis Group, London; New York, pp. 3–36 (2012). www.crcpress.com.

[7.9] C. G. Jin, T. Yu, Y. Yang, Z. F. Wu, L. J. Zhuge, X. M. Wu, and Z. C. Feng, "Ferromagnetic and photoluminescence properties of Cu-doped ZnO nanorods by radio frequency magnetron sputtering", Mater. Chem. Phys. **139**, 506–510 (2013). http://dx.doi.org/10.1016/j.matchemphys.2013.01.049.

[7.10] J. Chen, J. Zhang, J. Dai, F. Wu, S. Wang, H. Long, R. Liang, J. Xu, C. Chen, Z. Tang, Y. He, M. Li, and Z. C. Feng, "Significant anisotropic optical properties of heteroepitaxial strained nonpolar a-plane ZnO layers grown on $Al_xGa_{1-x}N$ templates", Opt. Mater. Exp. **7**, 3944–3951 (2017). https://doi.org/10.1364/OME.7.003944.

[7.11] J. Chen, J. Wang, N. Lu, V. Saravade, I. T. Ferguson, W. Hu, Z. C. Feng, and L. Wan, "Influences of Ni-doping on the structural and luminescent properties of ZnO thin films grown on sapphire by MOCVD", J. Vac. Sci. Technol. A **39**, 023402 (2021). https://doi.org/10.1116/6.0000816.

[7.12] A. Mondal, S. Pal, A. Sarkar, T. S. Bhattacharya, S. Pal, A. Singha, S. K. Ray, P. Kumar, D. Kanjilal, and D. Jana, "Raman investigation of N-implanted ZnO: Defects, disorder and recovery", J. Raman Spectrosc. **50**, 1926–1937 (2019). https://doi.org/10.1002/jrs.5732.

[7.13] Z. Mao, C. Fu, X. Pan, X. Chen, H. He, W. Wang, Y. Zeng, and Z. Ye, "Raman-based measurement of carrier concentration in n-type ZnO thin films under resonant conditions", Phys. Lett. A. **384**, 126148 (2020). https://doi.org/10.1016/j.physleta.2019.126148.

[7.14] R. L. de Sousa e Silva and A. Franco, Jr., "Raman spectroscopy study of structural disorder degree of ZnO ceramics", Mater. Sci. Semicond. Process. **119**, 105227 (2020). https://doi.org/10.1016/j.mssp.2020.105227.

[7.15] J. Villafuerte, O. Chaix-Pluchery, J. Kioseoglou, F. Donatini, E. Sarigiannidou, J. Pernot, and V. Consonni, "Engineering nitrogen- and hydrogen-related defects in ZnO nanowires using thermal annealing", Phys. Rev. Mater. **5**, 056001 (2021). https://doi.org/10.1103/PhysRevMaterials.5.056001.

[7.16] E. Przezdziecka, K. M. Paradowska, R. Jakiela, S. Kryvyi, E. Zielony, E. Placzek-Popko, W. Lisowski, P. Sybilski, D. Jarosz, A. Adhikari, M. Stachowicz, and A. Kozanecki, "Polar and non-polar $Zn_{1-x}Mg_xO$:Sb grown by MBE", Materials **15**, 8409 (2022). https://doi.org/10.3390/ma15238409.

[7.17] L. Ali, W. H. Shah, A. Ali, S. M. Eldin, A. A. Al-Jaafary, A. Sedky, J. Mazher, N. Imran, and M. Sohail, "Investigation of bulk magneto-resistance crossovers in iron doped zinc-oxide using spectroscopic techniques", Front. Mater. **10**, 1112798 (2023). https://doi.org/10.3389/fmats.2023.1112798.

[7.18] C.-C. Wang, A.-Y. Lo, M.-C. Cheng, Y.-S. Chang, H.-C. Shih, F.-S. Shieu, and H.-T. Tsai, "Zinc oxide nanostructures enhanced photoluminescence by carbon-black nanoparticles in Moire heterostructures", Sci. Rep. **13**, 9704 (2023). https://doi.org/10.1038/s41598-023-36847-1.

[7.19] D. R. Gutiérrez, G. García-Salgado, A. Coyopol, E. Rosendo-Andrés, R. Romano, C. Morales, A. Benítez, F. Severiano, A. M. Herrera, and F. Ramírez-González, "Effect of the deposit temperature of ZnO doped with Ni by HFCVD", Materials **16**, 1526 (2023). https://doi.org/10.3390/ma16041526.

[7.20] A. R. Bhapkar, M. Geetha, D. Jaspal, K. Gheisari, M. Laad, J.-J. Cabibihan, K. K. Sadasivuni, and S. Bhame, "Aluminum doped ZnO nanostructures for efficient photodegradation of indigo carmine and azo carmine G in solar irradiation", Appl. Nanosci. **13**, 5777–5793 (2023). https://doi.org/10.1007/s13204-023-02824-3.

[7.21] A. Muravitskaya, A. Rumyantseva, S. Kostcheev, V. Dzhagan, O. Stroyuk, and P.-M. Adam, "Enhanced Raman scattering of ZnO nanocrystals in the vicinity of gold and silver nanostructured surfaces", Opt. Exp. **24**, A168 (2016). https://doi.org/10.1364/OE.24.00A168.

[7.22] B. Hadzic, B. Matovic, M. Randjelovic, R. Kostic, M. Romcevic, J. Trajic, N. Paunovic, and N. Romcevic, "Phonons investigation of ZnO@ZnS core-shell nanostructures with active layer", J. Raman Spectrosc. **52**, 616–625 (2021). https://doi.org/10.1002/jrs.6058.

[7.23] M. Toma, O. Selyshchev, Y. Havryliuk, A. Pop, and D. R. T. Zahn, "Optical and structural characteristics of rare earth-doped ZnO nanocrystals prepared in colloidal solution", Photochem **2**, 515–527 (2022). https://doi.org/10.3390/photochem2030036.

[7.24] W. Yang, Z. Wu, Z. Liu, A. Pang, Y. L. Tu, and Z. C. Feng, "Room temperature deposition of Al-doped ZnO films on quartz substrates by radio-frequency magnetron sputtering and effects of thermal annealing", Thin Solid Films **519**, 31–36 (2010). https://doi.org/10.1016/j.tsf.2010.07.048.

[7.25] N. Ghobadi, M. Shiravand, and E. G. Hatam, "Influence of sputtered time on the structural and optical characterization of Al-doped ZnO thin films prepared by RF sputtering technique", Opt. Quant. Electron. **54**, 14 (2021). https://doi.org/10.1007/s11082-020-02687-w.

[7.26] C. Guillén and J. F. Trigo, "Ellipsometric study on the uniformity of Al:ZnO thin films deposited using DC sputtering at room temperature over large areas", Materials **16**, 6644 (2023). https://doi.org/10.3390/ma16206644.

CHAPTER 8

GaO-Based Oxides and Graphene

8.1 RAMAN SCATTERING OF β-Ga$_2$O$_3$ FILMS

We have performed Raman scattering studies on β-Ga$_2$O$_3$-based epitaxial and bulk materials [8.1–8.6]. Raman scattering has been recognized as an important and useful technical tool for intense and frontier research of gallium oxides and related materials [8.7–8.14]. The monoclinic gallium oxide (β-Ga$_2$O$_3$) with the wide bandgap (~4.9 eV) and very high breakdown field (8 MV/cm) is a thermally and chemically stable transparent conductive oxide and has been recently in hot and frontier research. It is considered a valuable material for novel devices ranging from deep ultraviolet (DUV) to infrared (IR) regions which include transparent field-effect transistors, DUV detectors, light-emitting diodes, gas sensors, microwaves, and optical masers. One must note that β-Ga$_2$O$_3$ makes it suitable for radiation resistance, high-frequency, and high-power electronics after SiC and GaN materials [8.1–8.14].

The atoms in gallium oxide can be arranged in several different forms known as polymorphs (i.e., β-, α- and ε-Ga$_2$O$_3$), with β-Ga$_2$O$_3$ form being the most stable polymorph. Figure 8.1 shows **a** schematic representation of the unit cell of β-Ga$_2$O$_3$. The tetrahedral coordination and octahedral coordination of Ga atoms are represented by black geometric structures, respectively [8.2].

Many techniques have been developed in recent years for synthesizing β-Ga$_2$O$_3$ thin films on semiconducting substrates. Recent methods used for preparing wide-bandgap oxides include magnetron sputtering, electron beam evaporation, metal-organic chemical vapor deposition (CVD), molecular beam epitaxy, sol–gel, and pulse laser deposition (PLD) [8.2, 8.6]. These methods offer versatile capabilities for growing heterostructures with novel material functionalities on different substrates including Si, Ge, GaAs, and sapphire (Al$_2$O$_3$) [8.2].

For our investigation, a set of beta gallium oxide (β-Ga$_2$O$_3$) ultrathin films were grown on sapphire with different growth temperatures (T$_g$ = 400–1000 °C) by pulsed laser deposition, with thicknesses between 75 and 1100 nm. We have performed comprehensive

DOI: 10.1201/9781032644912-8

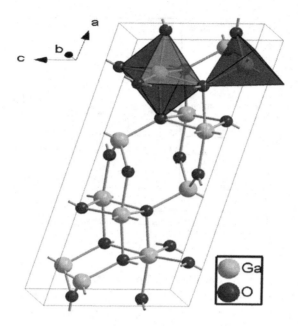

FIGURE 8.1   Schematic representation of the unit cell of β-Ga_2O_3. The tetrahedral coordination and octahedral coordination of Ga atoms are represented by black geometric structures, respectively.

Source: From [8.2], figure 1, with reproduction permission of Elsevier.

structural, electrical, and optical studies on these β-Ga_2O_3, via X-ray absorption spectroscopy, Raman scattering, and X-ray photoelectron spectroscopy (XPS) [8.2]. Figure 8.2 presents room-temperature (RT) Raman spectra, excited under 532 nm of five Ga_2O_3 films grown on sapphire.

Several sharp vibrational peaks at 379, 419, 431, 449, 576, and 750 cm^{-1} are phonon modes from the sapphire substrate. Raman-active modes from β-Ga_2O_3 are observed at 202 and 346 cm^{-1} appeared only in three samples of S3, S5b, and S5c. The effect of higher deposition temperatures in the Ga_2O_3 films can be seen obviously from both the low-frequency (202 cm^{-1}) and mid-frequency (346 cm^{-1}) peaks. They belong to the tetrahedral–octahedral mode (202 cm^{-1}) of GaO_4 and the octahedral mode (346 cm^{-1}) of GaO_6. On the other hand, the samples S1 (T_g = 400°C) and S2 (T_g = 550°C) showed no β-Ga_2O_3 modes, indicating that Ga_2O_3 films might not be formed perfectly under the low deposition temperatures.

The β-Ga_2O_3 Raman-active peaks at 202 cm^{-1} are enlarged at the upper-right corner (experimental data in dots) in Figure 8.2. They are theoretically fitted by the spatial correlation model (SCM), using Eqs. (1.16–1.18), with fitting results displayed by black solid lines in the insert graph. Our fitting results showed that sample S3 possesses the narrowest Raman bandwidth Γ and largest phonon propagation length L while sample S5b has the second narrower Raman line width Γ and second-larger phonon propagation length L. Obviously, sample S3 (T_g = 700°C) can be identified as the best-quality sample, while S5b (T_g = 1000°C) as second and S5c (also T_g = 1000°C) at third. Raman scattering has proved

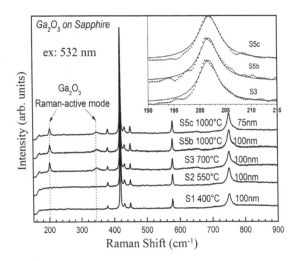

FIGURE 8.2   Raman spectra at RT and under 532 nm excitation of five Ga_2O_3 films grown on sapphire, showing Raman-active modes from β-Ga_2O_3 at 202 and 346 cm^{-1}, and other peaks from sapphire substrate. The upper right corner is an enlarged view (experimental data in dots) and fitting results (black solid lines) of β-Ga_2O_3 Raman-active peak at 202 cm^{-1}.

Source: **From [8.2], figure 3, with reproduction permission of Elsevier.**

the crystalline quality of this set of Ga_2O_3 films by analyzing the Raman line shape via the spatial correlation modeling process [8.2].

8.2   RAMAN SCATTERING OF FE- AND TA-DOPED β-Ga_2O_3 CRYSTALS

There is a great research interest in β-Ga_2O_3 crystals doped with trivalent ions, which have grown rapidly because of their technological significance in the development of various optical and microwave devices, as well as in the study of the exchange interactions between substitution ions. Fe is to serve as an effective capture center in a variety of group-II and group-III semiconductors. The added Fe iron in β-Ga_2O_3 crystal has the same valence state (3+) as the Ga^{3+} ion, with its ion radius (0.064 nm) very close to the Ga^{3+} (0.062 nm) ion radius. Therefore, Fe^{3+} can be easily substituted at the Ga site without supporting the lattice with an electron source (donor) or (acceptor), but it acts as a trap level or capture center for the excited carriers to compensate for unintentional n-type conductivity. Fe^{3+} substitution at Ga sites has been shown to be a strongly favorable capture center in β-Ga_2O_3 crystals, which dominate deep defect traps [8.4 and references therein]. Our research team has prepared Fe-doped β-Ga_2O_3 crystals by the edge-defined film-fed growth method and performed a systematic investigation of their optical and electronic band properties, before and after annealing, by using a combination of high-resolution X-ray diffraction (HR-XRD), X-ray and ultraviolet photoelectron spectroscopy (XPS/UPS), Raman scattering, optical transmission, and electron paramagnetic resonance spectroscopy techniques [8.4].

Figure 8.3 shows the RT Raman spectra, using a 532-nm excitation laser source (Horiba iHR 550 spectrometer) and excited under 532 nm of undoped and 0.05 mol% Fe-doped β-Ga_2O_3 crystals, as well as the effect of air-annealing treatment [8.4]. Lorentz fits were

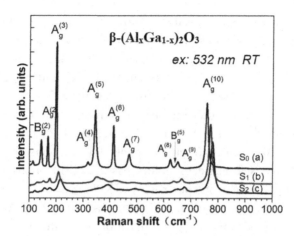

FIGURE 8.3   Raman spectra of the undoped and 0.05 mol% Fe-doped β-Ga_2O_3 single crystals as-grown and air-annealed, measured at RT and 532 nm excitation.

Source: **From [8.4], figure 2, with reproduction permission of Springer.**

applied to obtain more accurate information on experimental Raman peaks and the line-width. Eleven Raman peaks are detected from undoped and Fe-doped β-Ga_2O_3 crystals. The peaks observed at 115, 146, 171, 201, 321, 348, 418, 477, 632, 660, and 768 cm^{-1} correspond to the A(1) g, B(2) g, A(2) g, A(3) g, A(4) g, A(5) g, A(6) g, A(7) g, A(8) g, A(9) g, and A(10) g phonon modes, respectively. The low-frequency phonon modes (B(2) g, A(2) g, A(3) g) at 145, 170, and 201 cm^{-1} are attributed to the liberation and translations of tetrahedral–octahedral chains, which show a slight change upon adding Fe. This slight change could be due to two competing mechanisms: (a) the softening mode arises from the increase in the lattice parameters due to Fe and (b) the hardening mode arises because of the reduction in the effective atomic mass (by adding Fe). The mid-frequency phonon modes (A(4) g, A(5) g, A(6) g, A(7) g) at 321, 348, 417, and 476 cm^{-1} are ascribed to the deformation of the GaO_4 tetrahedron and GaO_6 octahedron [8.4–8.6, 8.13]. There was an increase in the intensity of the A(4) g and A(7) g g modes and a decrease in the A(5) g and A(6) g modes after adding Fe.

The high-frequency phonon modes (A(8) g, A(9) g, A(10) g) at 631, 660, and 768 cm^{-1} are related to the stretching and bending of the tetrahedron [8.4, 8.6, 8.14], which were inhibited by the added Fe atoms. A significant decrease in A(10) g can be observed, indicating that Fe atoms mainly affect the symmetry of the tetrahedral site. A slight shift is observed in the phonon frequency modes after the addition of Fe. This is attributed to the partial substitution of the small Ga^{3+} ion with the large Fe^{3+} ion, leading to an expansion of the lattice and an increase in the bond length. An increase in the bond length corresponds to a decrease in the phonon frequency, which is revealed by the slight decreases in the high-frequency phonon modes (A(8) g, A(9) g, A(10) g) after adding Fe. Our Raman scattering studies revealed that the high-frequency phonon modes that belong to the stretching and bending of Ga_1O_4 tetrahedral were significantly inhibited by the Fe addition to the β-Ga_2O_3 crystal [8.4].

Also, our research team has prepared 0.10 mol% Ta-doped β-Ga_2O_3 crystal using the optical floating zone method. The starting materials are β-Ga_2O_3 (6N) and Ta_2O_5 (4N) powder. After pressing the bar with a cold isostatic press, it was sintered in a high-temperature furnace for 20 hours in an air environment at 1450°C, and a <010> oriented crystal was used as the seed. Growth was carried out using Quantum Design IRF01–001–00 infrared image furnace. The sintered rods and seed were rotated at 10 rpm in opposite directions, and the crystals were grown in flowing air at the speed of 5 mm/h. After growth, the sample was cut into $6 \times 10 \times 1$ mm³ samples for annealing and polishing. They were investigated by HR-XRD, Raman scattering, photoluminescence (PL), photoluminescence excitation (PLE), and X-ray photoelectron spectroscopy (XPS) [8.5]. Raman spectra showed ten Raman-active modes of B(2) g, A(2) g, A(3) g, A(4) g, A(5) g, A(6) g, A(7) g, A(8) g, A(9) g, and A(10) g, attributed to the vibration and translation of tetrahedra-octahedra chains and associated to the deformation of GaO_6 octahedra, as well as the stretching and bending of GaO_4 tetrahedra, respectively. It is observed that most of the Raman peaks exhibit blue shifts and their intensities increase except the A10 g mode. This indicates an expansion of the crystal lattice and an improvement in crystal quality. In addition, the A(10) g mode related to GaO_4 tetrahedra is reduced by annealing treatment indicating that the symmetry of the gallium oxide GaO_4 tetrahedron was weakened in this site [8.5].

8.3 RAMAN SCATTERING AND TEMPERATURE DEPENDENCE OF β-$(Al_xGa_{1-x})_2O_3$ CRYSTALS

Bulk crystals of β-$(Al_xGa_{1-x})_2O_3$ with x(Al) = 0, 0.1 and 0.2, that is, β-Ga_2O_3, β-$(Al_{0.1}Ga_{0.9})_2O_3$ and β-$(Al_{0.2}Ga_{0.8})_2O_3$, were grown by the optical floating zone and edge-defined film-fed methods. Their microstructure and temperature-dependent phonon characteristics were investigated systematically using HR-XRD and Raman scattering spectroscopy from our research team [8.3]. The HR-XRD results revealed a very good crystalline phase for all samples. As compared to the pure β-Ga_2O_3, the interplanar spacings in $(Al_{0.1}Ga_{0.9})_2O_3$ and $(Al_{0.2}Ga_{0.8})_2O_3$ samples are found decreasing with the increase of Al contents, indicating distorted lattice structures. Figure 8.4 shows Raman spectra, measured at RT and under the 532 nm excitation of S_0, S_1, and S_2, that is, β-$(Al_xGa_{1-x})_2O_3$ with x(Al) = 0, 0.1 and 0.2, respectively [8.3].

The Raman-active phonon modes are closely related to the symmetry of the crystal structure. In the monoclinic structure of β-Ga_2O_3 crystal – the two $[GaO_6]$ octahedrons are connected with two $[GaO_4]$ tetrahedrons to form double chains along the b axis. Each β-Ga_2O_3 primitive cell consists of 10 atoms, generating 30 phonon modes, 27 of which are the optical phonons. These optical modes at the center of Brillouin can be expressed as [8.3, 8.6, 8.7, 8.11, 8.14]:

$$\Gamma^{opt} = 10A_g + 5B_g + 4A_u + 8B_u \tag{8.1}$$

where A_g and B_g are symmetrical Raman-active phonon modes, while the A_u and B_u are infrared active phonon modes. As mentioned in Section 8.2, the Raman-active phonon modes of these β-Ga_2O_3 bulk crystals appear in low-frequency region (B(2) g, A(2) g, A(3) g),

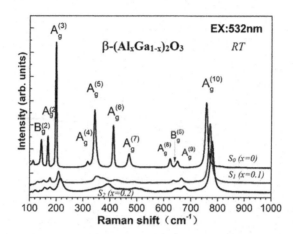

FIGURE 8.4 RT Raman scattering spectra of S_0, S_1, and S_2, measured at RT and under the 532 nm excitation.

Source: From [8.3], figure 3, with reproduction permission from Elsevier.

mid-frequency (A(4) g, A(5) g, A(6) g, A(7) g), and high-frequency (A(8) g, A(9) g, A(10) g), which belong to the liberation and translation of the [GaO$_4$] chain, the deformation of [GaO$_4$] tetrahedrons and [GaO$_6$] octahedrons, the stretching and bending of [GaO$_4$] tetrahedrons, respectively [8.4, 8.6–8.8, 8.11–8.14].

Raman peaks of $(Al_xGa_{1-x})_2O_3$ in Figure 8.4 exhibited blueshifts and linewidth broadening with an increase of x. The lattice distortion mainly affected the Raman modes below 600 cm^{-1} and had a relatively trivial effect on the modes higher than 600 cm^{-1}. As compared to the pure bulk β-Ga$_2$O$_3$, the Raman peaks of $(Al_xGa_{1-x})_2O_3$ samples S_1 and S_2 have changed significantly. Except for A$_g$(10), most of the Raman peaks and their intensities are weakened with full widths at half maxima (FWHMs) broadened. In the Raman spectrum of pure β-Ga$_2$O$_3$, the strongest mode is A$_g^{(3)}$, while in the spectra of two $(Al_xGa_{1-x})_2O_3$ samples S_1 and S_2, the strongest peak is A$_g^{(10)}$. It is also noticed that the shapes of Raman peaks of S_1 and S_2 in the medium wavenumber region become much distorted and no longer remain as regular Lorentzian-shaped peaks. Their linewidths are broadened significantly, more than three times larger than those of the pure sample. Obviously, the lattice distortion with Al atoms replacing Ga atoms destroys the symmetry of the crystal structure and weakens the Raman-active modes. Moreover, Al-alloying has a great influence on the Raman modes in the region below 600 cm^{-1}, that is, on the deformation of [GaO$_4$] tetrahedrons and [GaO$_6$] octahedrons. As the Al atom tends to replace Ga in octahedral sites for $(Al_xGa_{1-x})_2O_3$ crystals, [GaO$_4$] tetrahedrons are less distorted than those of [GaO$_6$] octahedrons. The vibrational modes in high wavenumber region (> 600 cm^{-1}) mainly from [GaO$_4$] tetrahedrons are slightly affected, especially for the Raman mode A$_g^{(10)}$ [8.3, 8.4, 8.6–8.8, 8.11–8.14].

Figure 8.5 presents temperature-dependent Raman Spectra of S_0, S_1, and S_2 in 80–800 K, respectively. With the increase of temperature (T), the Raman peaks are redshifted and their linewidth FWHMs broadened. Like the pure bulk β-Ga$_2$O$_3$, Raman shifts in the two

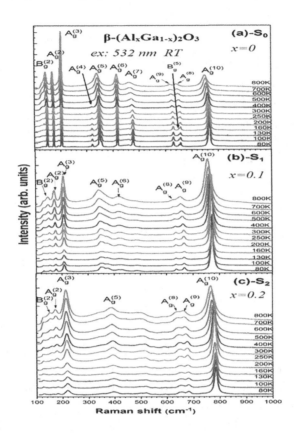

FIGURE 8.5 Variable temperature (80–800 K) Raman spectra of three $(Al_xGa_{1-x})_2O_3$ with x = 0, 0.1 and 0.2, respectively.

Source: From [8.3], figure 4, with reproduction permission from Elsevier.

$(Al_xGa_{1-x})_2O_3$ crystals revealed quadratic dependence on temperature. But the T-dependent widening of Raman-active phonons in $(Al_xGa_{1-x})_2O_3$ exhibited a convoluted behavior different from β-Ga_2O_3 [8.3]. In the variable temperature (VT) Raman scattering spectra of S_0, S_1, and S_2 samples, with the increase of T, the Raman peaks are redshifted. At higher temperatures, some modes are broadened and overlapped with each other, such as $A_g^{(4)}$ and $A_g^{(5)}$, $B_g^{(5)}$ and $A_g^{(9)}$ for the pure sample (S_0), and $A_g^{(8)}$ and $A_g^{(9)}$ for two $(Al_xGa_{1-x})_2O_3$ samples (S_1, S_2). At lower temperatures below 200 K, the intensities of other modes are greatly reduced and almost disappeared (e.g., $B_g^{(2)}$ and $A_g^{(2)}$) for $(Al_{0.2}Ga_{0.8})_2O_3$ sample (S_2)). As the temperature increased, the lattice vibrations were strengthened, and the intensities of most Raman modes increased. On the other hand, with the expansion of the crystal lattice, the restoring force of the lattice vibration decreased, and the Raman peaks redshifted [8.3, 8.11–8.14].

The temperature dependence of Raman phonon shifts for the three samples displayed in Figure 8.5 can be fitted and analyzed by using Eqs. (1.24–1.27). There are contributions from linear thermal expansion, anharmonic phonon coupling, three-phonon process, and four-phonon process, respectively. The Raman scattering measurements have found Raman-active modes redshifted and their FWHMs widened with Al-alloying. The Raman-active modes in the medium and low wavenumber regions (<600 cm^{-1}) were greatly weakened, indicating that the Al doping mainly affected the deformation of [GaO$_4$] tetrahedrons and [GaO$_6$] octahedrons, and the liberation and translation of the [GaO$_4$] chain. The decrease of Raman shifts of optical phonons with increasing temperature is mainly due to the change of harmonic frequency with volume expansion and the anharmonic coupling to phonons of other branches. For the Al$_x$Ga$_{1-x}$)$_2$O$_3$ samples, some Raman modes are broadened and merged at high temperatures, while the others are weakened and disappear at low temperatures. Our results revealed the lattice distortion and change of crystal symmetry and indicated that the Al-alloying induced complex lattice vibrations and phonon coupling [8.3].

8.4 SPATIAL CORRELATION AND ANISOTROPY OF β-(Al$_x$Ga$_{1-x}$)$_2$O$_3$ CRYSTALS

Another set of β-(Al$_x$Ga$_{1-x}$)$_2$O$_3$ crystals with five different Al compositions (x = 0.0, 0.06, 0.11, 0.17, 0.26) were prepared by optical floating zone method, and systematically studied by our team via SCM and the angle-resolved polarized Raman spectroscopy for the long-range crystallographic order and anisotropy [8.6]. Raman spectra of all five samples have revealed 10 Raman-active phonon features (B2 g, A2 g, A3 g, A4 g, A5 g, A6 g, A7 g, A8 g, A9 g, and A10 g) in the frequency range of 100–1000 cm^{-1}, like the cases discussed in Sections 8.2–8.3 and references [8.3–8.5, 8.8, 8.11–8.14]. Alloying with aluminum is seen as causing Raman peaks to blue shift while their FWHMs broadened.

The SCM (Eqs. (1.14–1.16)) of Raman scattering spectra has been used to analyze the effects of Al composition, x, on the lattice vibrational modes. The correlation length (CL) decreases with the increase of Al composition, indicating a decrease in the orderliness of the lattice vibrations due to the Al doping. The relationship between the CL and the Al composition is fitted linearly. The phonons in the high-frequency region are only affected by the atomic substitution in the sublattice related to their vibration, resulting in relatively smaller changes in the spatial correlation.

In a perfect crystal with translational symmetry, the spatial correlation function of phonons is infinite, following the selection rule of q = 0. As the aluminum composition is increased in our β-(Al$_x$Ga$_{1-x}$)$_2$O$_3$ samples, each Raman peak produces different degrees of asymmetric broadening and causes a blue shift, the disorder of microstructures increases and the CL of the Raman modes decreases. As x increases, the CL is more affected for low-frequency phonons than the modes in the high-frequency region. For each Raman mode, the CL is decreased with increasing temperature.

Angle-resolved polarized Raman spectroscopy (ARPRS) has been utilized to investigate the anisotropic properties of various alloyed β-(Al$_x$Ga$_{1-x}$)$_2$O$_3$ crystals. The ARPRS spectra for all five β-(Al$_x$Ga$_{1-x}$)$_2$O$_3$ (x = 0.0, 0.06, 0.11, 0.17, 0.26) samples were measured in parallel polarization and cross-polarization configurations and displayed in Waterfall plots [8.6]. Figure 8.6 shows two typical Waterfall plots of S$_2$, β-(Al$_x$Ga$_{1-x}$)$_2$O$_3$ with x = 0.11, measured at RT, under 532 nm excitation, and in parallel polarization (a) and cross-polarization (b) configurations, with the B2 g, A3 g, A5 g, and A10 g peaks indicated.

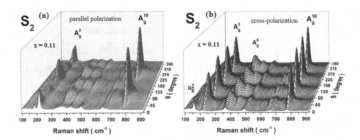

FIGURE 8.6   Waterfall plots of the angle-resolved polarized Raman spectra for S_2, β-$(Al_xGa_{1-x})_2O_3$ with x = 0.11, measured at RT and under 532 nm excitation, in (a) parallel polarization and (b) cross-polarization configurations.

Source: From [8.6], figure 7, with reproduction permission from the open-access article of MDPI Materials.

Like Eqs. (3.3–3.5) for GaN, the ARPRS intensity expression in the plane (100) for the A_g and B_g modes of gallium oxide can be obtained as follows:

$$I_{A_g}^{\parallel} \propto |a|^2 \sin^4\theta + |c|^2 \cos^4\theta + 2|a||c|\sin^2\theta\cos^2\theta\cos_{\varphi_{ac}} \tag{8.2}$$

$$I_{A_g}^{\perp} \propto \sin^2\theta\cos^2\theta\left(|a|^2 + |c|^2 - 2|a||c|\cos_{\varphi_{ac}}\right) \tag{8.3}$$

$$I_{B_g}^{\parallel} \propto \left[|e|\sin(2\theta)\right]^2 \tag{8.4}$$

$$I_{B_g}^{\perp\parallel} \propto \left[|e|\cos(2\theta)\right]^2 \tag{8.5}$$

where a, c, and e are the independent Raman tensor elements, θ is the angle between the polarization direction of the incident light and the b axis of the sample, φ_{ac} is the phase difference between the two distinct elements of the Raman tensor.

To figure out the differential polarization degree of the internal structure of the aluminum alloy GaO sample, Eqs. (8.2) and (8.3) are used to fit the polar coordinates of each mode. The fitting curves are consistent with the experimental data. The anisotropic properties of β-$(Al_xGa_{1-x})_2O_3$ alloy samples on the (100) plane can be analyzed through the ratio of Raman tensor elements |a/c| and the phase difference φ. For the A_g mode, the value of |a/c| reflects the anisotropy of the Raman tensor. Relatively, A10 g at ~774 cm^{-1} exhibits strong anisotropy. The A10 g mode involves only the stretching of the two Ga–O bonds, resulting in a strong anisotropy on the (100) plane. The average bond length of Al–O is shorter than that of Ga–O. When Al atoms replace Ga atoms, the polarization of the original Ga–O bonds in the (100) plane is decreased, and their anisotropy is weakened. The anisotropy of the Raman phonon modes of β-Ga_2O_3 is strongly influenced by aluminum alloying. Obviously, the anisotropy of A3 g increases with the increase of aluminum content, while the incorporation of aluminum weakens the anisotropy of A10 g. This may be caused by the large difference in the bond polarizability. The A3 g phonon is mainly affected by chain oscillation, and the vibration direction of the involved atoms is more diverse, resulting in a more significant anisotropy [8.6]. Our comprehensive study has provided meaningful results for comprehending the long-range orderliness and anisotropy in technologically important β-$(Al_xGa_{1-x})_2O_3$ crystals.

8.5 RAMAN SCATTERING OF GRAPHENE ON Si AND SiO$_2$

Since the experimental realization of one-atom-thick graphene sheets along with the measurements of the quantum Hall effect, a great deal of interest has emerged in both the fundamental research and the development of device engineering concepts. Graphene has an extremely high carrier mobility of ~15,000 cm^2 V^{-1} S^{-1} and thermal conductivity of 5000 W m^{-1} K^{-1} with a very strong Young's modulus of ~1 TPa. The Dirac Fermions in graphene have caused both integer and fractional quantum Hall effects. Unconventional superconductivity has also been realized in a two-dimensional superlattice created by stacking two sheets of twisted graphene relative to each other by a small angle. Along with the unique electronic features, graphene has displayed extraordinary optical responses. Graphene, being a one-atom-thick sheet of carbon, exhibits significant absorption in the visible to infrared wavelength region (2.3%) with reflectance less than 0.1%. This indicates that a one-atom-thick graphene layer is extremely transparent, having a high degree of flexibility with excellent optical properties. A variety of graphene commercial applications have been intensified and in wide developments [8.15–8.22 and references therein].

We have performed Raman scattering investigation on graphene materials [8.15, 8.16]. Raman scattering has been employed as an important and useful technical tool for intense and frontier research of graphene [8.17–8.22]. Figure 8.7 presents Raman spectra of four graphene layers on Si and SiO$_2$/Si substrate. Graphene was first prepared on copper foil by the CVD method and then transferred onto Si and on SiO$_2$/Si substrates. For each type of substrate, both monolayer and bilayer graphene were prepared. The graphene thicknesses were determined by spectroscopic ellipsometry measurements as 0.38 and 0.73 nm for one-layer and two-layer graphene/Si, 0.34 nm and 0.88 nm for one-layer and two-layer graphene/SiO$_2$/Si, respectively [8.15].

Raman spectroscopy was employed by us to characterize the quality of the transferred graphene. As shown in Figure 8.7, all graphene samples exhibit two intrinsic Raman peaks (G, 2D). Two disorder-induced peaks are displayed with (D, D+D") peaks for graphene on Si, and (D, D'+D") peaks for graphene on SiO$_2$/Si. The defect-activated peaks, D'+D" and D+D", correspond to the combination mode of the D' and D" modes as well as the D and D" modes. The intensity of Raman scattering light for the graphene on Si substrate is weaker than that of graphene on SiO$_2$/Si substrate.

The Raman scattering results indicated that there existed a degree of defects in the four transferred graphene sheets. The intensity ratio between the 2D band and the G band can determine the number of graphene layers. The I_{2D}/I_G values higher than 2, between 1 and 2, and lower than 1, correspond to the presence of monolayer graphene, bilayer graphene, and three or more layers, respectively. As for our four graphene samples, the I_{2D}/I_G values of two monolayer graphene were respectively 2.1 (on Si substrate) and 3.4 (on SiO$_2$/Si substrate) while those of two bilayer graphene were 1.5 (on Si substrate) and 1.7 (SiO$_2$/Si substrate), respectively [8.15].

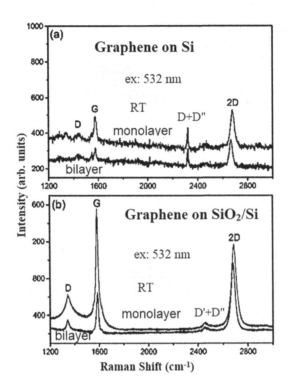

FIGURE 8.7   Raman scattering spectra at RT and under the 532 nm excitation of graphene samples of (a) graphene on Si substrate and (b) graphene on SiO$_2$/Si substrate.

Source: **From [8.15], figure 1, with reuse permission from the open-access article of MDPI Crystals**

8.6 RAMAN SCATTERING OF GRAPHENE ON SiC AND TEMPERATURE DEPENDENCE

Our research team has conducted a systemic investigation to explore structural, optical, and temperature-dependent properties of single- and bilayer graphene on 4H–SiC substrates for triboelectric nanogenerators (TENGs). TENGs can convert mechanical energy into electrical energy for power supply without polluting the environment. As an emerging distributed energy technology, TENGs are widely used in various fields such as renewable energy, electronic information, communications, and health care. To improve the performance of TENG, materials that are prone to charge transfer are highly desirable. Monolayer graphene is an ultra-thin two-dimensional material and the distance between C–C is about 0.142 nm, owning unique mechanical, electrical, and optical properties. Its excellent properties have promoted the development of electronic devices and sensor devices. Graphene has excellent properties such as ultra-high electrical conductivity, high carrier mobility, and thermal conductivity, with a promising application in the field of TENGs [8.16 and references therein].

In our studies [8.16], by the transferred graphene methods, the single-layer and bilayer graphene were first grown on the copper foil by the CVD method and subsequently

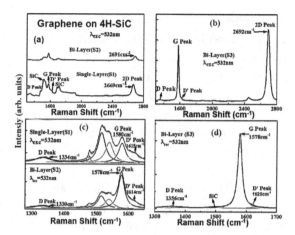

FIGURE 8.8 (a)–(b) Room-temperature Raman spectra measured under 532 nm of graphene samples (S1), (S2), and (S3), (c)–(d) for the Lorentz fittings, respectively.

Source: **From [8.16], figure 3, with reuse License (Permission) Grant from Frontiers for an open-access article.**

transferred to 4H–SiC substrate to obtain SiC-based graphene samples, labeled S1 and S2, respectively. Also, by the decomposition of the SiC method, a bilayer graphene was prepared on the Si-terminated 4H–SiC as sample S3. The size of the experimental samples is 1×1 cm². Figure 8.8(a) and (b) show the RT Raman spectra of the graphene on 4H–SiC samples S1 (single layer), S2 (bilayer), and S3 (bilayer), respectively. Figure 8.8(c) and (d) present the Lorentz fittings of the Raman peaks.

By Lorentzian fit in Figure 8.8(c), the G peak of sample S1 is located at 1580 cm⁻¹, with the FWHM of 28.5 cm⁻¹, and the 2D peak is at 2669 cm⁻¹. The G peak of the sample S2 is at 1578 cm⁻¹ with a FWHM of 23.7 cm⁻¹, and the 2D peak of S2 is at 2691 cm⁻¹. As seen in Figure 8.8(d), the sample S3 G peak is at 1578 cm⁻¹ with a FWHM of 16.4 cm⁻¹, and its 2D peak is at 2692 cm⁻¹. In addition, the D peak FWHM of S3 is 69.7 cm⁻¹, while the FWHM (D) of S1 and S2 is 55 and 106.1 cm⁻¹, respectively, indicating that the transferred graphene has more defects. It is worth noting that the 2D peak of sample S3 has a blue shift of about 1 cm⁻¹ compared with sample S2, and about 23 cm⁻¹ compared with sample S1, which would be attributed to the strong force between the SiC substrate and the graphene.

Generally, I_G/I_{2D} can be used to judge the number of graphene layers, and the intensity ratio of D peak to G peak (I_D/I_G) can reflect the quality (or grain size) of graphene film [8.15–8.22]. The D peak and the D′ peak are induced by the defects. The intensity ratio ($I_D/I_{D'}$) is related to defect density n_d, reflecting the defect level of graphene. Within a specific defect concentration range, the intensity ratio of D peak and D″ peak is proportional to the defect concentration ($I_D/I_{D'} \propto n_d$). The distribution and density of defects in graphene are important factors controlling the carrier transport properties of graphene. However, beyond a certain range of defect concentration, the intensity of the D peak will decrease, but the intensity of D′ peak will remain unchanged. In addition, the analysis of the ratio range of $I_D/I_{D'}$ indicates that there are more defects in the edge type of graphene.

The VT properties of graphene on SiC substrate are studied further from Raman measurements, including the temperature characteristics such as stress–strain and carrier concentration of graphene. Graphene and SiC have opposite thermal expansion coefficients. Within a specific temperature range, as the temperature rises, graphene exhibits a negative thermal expansion coefficient and shrinks, while SiC exhibits a positive thermal expansion coefficient and expands. Therefore, the graphene shrinks inward when the temperature rises, while the SiC substrate expands outward, leading to a compressive force on the graphene. Generally, both the G and 2D peaks are shifted downward in frequency with the increase in temperature.

The first-order Raman scattering peak G of graphene can be quantitatively analyzed from the SCM using Eqs. (1.14–1.16). Based on the spatial correlation theory, the phonon CL is used to measure the average distance between point defects to describe the crystal quality. Results showed that, with the increase in temperature, the CL gradually increases, and the ratio of Raman intensity (I_D/I_G) between D and G peaks decreases gradually. Compared to the transferred graphene onto SiC substrate, graphene grown by decomposition of SiC has better crystalline quality, fewer impurities, and a more stress effect between graphene and substrate, as well as better performance in TENG applications [8.16].

REFERENCES

[8.1] P. F. Huang, Y. T. Chen, H. Y. Lee, Z. C. Feng, H. H. Lin, and W. Lu, "Surface and material characteristics of Ga_2O_3 thin films on GaAs", Proc. SPIE **7067**, 70670M-1–8 (2008). https://doi.org/10.1117/12.795535.

[8.2] H. Yang, Y. Qian, C. Zhang, D.-S. Wuu, D. N. Talwar, H.-H. Lin, J.-F. Lee, L. Wan, K. He, and Z. C. Feng, "Surface/structural characteristics and band alignments of thin Ga_2O_3 films grown on sapphire by pulse laser deposition", Appl. Surf. Sci. **479**, 1246–1253 (2019.02). https://doi.org/10.1016/j.apsusc.2019.02.069.

[8.3] X. Chen, W. Niu, L. Wan, C. Xia, H. Cui, and Z. C. Feng, "Structure and temperature-dependence of Raman scattering properties of β-$(Al_xGa_{1-x})_2O_3$ crystals", Superlattices Microstruct. **140**, 106469 (2020). https://doi.org/10.1016/j.spmi.2020.106469.

[8.4] N. Zhang, H. Liu, Q. Sai, C. Shao, H. F. Mohamed, C. Xia, L. Wan, and Z. C. Feng, "Structural and electronic characterizations of Fe doped β-Ga_2O_3 single crystal", J. Mater. Sci. **56**, 13178–13189 (2021). https://doi.org/10.1007/s10853-021-06027-5.

[8.5] H. Liu, N. Zhang, J. Yin, C. Xia, Z. C. Feng, K. He, L. Wan, and H. F. Mohamed, "Characterization of defect levels in β-Ga_2O_3 single crystals doped with tantalum", CrystEngComm **23**, 2835–2841 (2021). https://doi.org/10.1039/D0CE01639J.

[8.6] L. Li, L. Wan, C. Xia, Q. Sai, D. N. Talwar, Z. C. Feng, H. Liu, J. Jiang, and P. Li, "The spatial correlation and anisotropy of β-$(Al_xGa_{1-x})_2O_3$ single crystal", Materials **16**, 4299 (2023). https://doi.org/10.3390/ma16124269.

[8.7] B. M. Janzen, P. Mazzolini, R. Gillen, A. Falkenstein, M. Martin, H. Tornatzky, J. Maultzsch, O. Bierwagen, and M. R. Wagner, "Isotopic study of Raman active phonon modes in β-Ga_2O_3", J. Mater. Chem. C **9**, 2311 (2021). https://doi.org/10.1039/D0TC04101G.

[8.8] K. Zhang, Z. Xu, J. Zhao, H. Wang, J. Hao, S. Zhang, H. Cheng, and B. Dong, "Temperature-dependent Raman and photoluminescence of β-Ga_2O_3 doped with shallow donors and deep acceptors impurities", J. Alloys Compd. **881**, 160665 (2021). https://doi.org/10.1016/j.jallcom.2021.160665.

[8.9] C. Remple, J. Huso, and M. D. McCluskey, "Photoluminescence and Raman mapping of β-Ga_2O_3", AIP Adv. **11**, 105006 (2021). https://doi.org/10.1063/5.0065618.

[8.10] B. M. Janzen, R. Gillen, Z. Galazka, J. Maultzsch, and M. R. Wagner, "First- and second-order Raman spectroscopy of monoclinic β-Ga$_2$O", Phys. Rev. Mater. **6**, 054601 (2022). https://doi.org/10.1103/PhysRevMaterials.6.054601.

[8.11] K. Zhang, Z. Xu, S. Zhang, H. Wang, H. Cheng, J. Hao, J. Wu, and F. Fang, "Raman and photoluminescence properties of un-/ion-doped β-Ga$_2$O$_3$ single-crystals prepared by edge-defined film-fed growth method", Phy. B Condens. Matter **600**, 412624 (2021). https://doi.org/10.1016/j.physb.2020.412624.

[8.12] K. V. Akshita, D. Dhanabalan, R. Hariharan, and S. M. Babu, "Effect of gamma-irradiation on structural, morphological, and optical properties of β-Ga$_2$O$_3$ single crystals", J. Mater. Sci. Mater. Electron. **34**, 841 (2023). https://doi.org/10.1007/s10854-023-10228-w.

[8.13] D. Das, G. Gutierrez, and C. V. Ramana, "Raman spectroscopic characterization of chemical bonding and phase segregation in Tin (Sn)-incorporated Ga$_2$O$_3$", ACS Omega **8**, 11709–11716 (2023). https://doi.org/10.1021/acsomega.2c05047.

[8.14] J. Bhattacharjee and S. D. Singh, "Observation of mixed-mode behavior of Raman active phonon modes for β-(Al$_x$Ga$_{1-x}$)$_2$O$_3$ alloys", Appl. Phys. Lett. **122**, 112101 (2023). https://doi.org/10.1063/5.0137855.

[8.15] S. Wu, L. Wan, L. Wei, D. N. Talwar, K. He, and Z. C. Feng, "Absorption, dispersion and temperature-dependent properties of graphene on Si and SiO$_2$ substrates", Crystals **11**, 358 (2021.3). https://doi.org/10.3390/cryst11040358.

[8.16] S. Wang, L. Wan, D. Li, X. Chen, X. Xu, Z. C. Feng, and I. T. Ferguson, "Temperature-dependent properties of graphene on SiC substrate for triboelectric nanogenerators", Front. Mater. **9**, 924143 (2022). https://doi.org/10.3389/fmats.2022.924143.

[8.17] K. PieRtak, J. Jagiełło, A. Dobrowolski, R. Budzich, A. Wysmołek, and T. Ciuk, "Enhancement of graphene-related and substrate-related Raman modes through dielectric layer deposition", Appl. Phys. Lett. **120**, 063105 (2022). https://doi.org/10.1063/5.0082694.

[8.18] X. Chen, S. Reichardt, M.-L. Lin, Y.-C. Leng, Y. Lu, H. Wu, R. Mei, L. Wirtz, X. Zhang, A. C. Ferrari, and P.-H. Tan, "Control of Raman scattering quantum interference pathways in graphene", ACS Nano **17**, 5956–5962 (2023). https://doi.org/10.1021/acsnano.3c00180.

[8.19] Z. Li, L. Deng, I. A. Kinloch, and R. J. Young, "Raman spectroscopy of carbon materials and their composites: Graphene, nanotubes and fibres", Prog. Mater. Sci. **135**, 101089 (2023). https://doi.org/10.1016/j.pmatsci.2023.101089.

[8.20] J. Sonntag, K. Watanabe, T. Taniguchi, B. Beschoten, and C. Stampfer, "Charge carrier density dependent Raman spectra of graphene encapsulated in hexagonal boron nitride", Phys. Rev. B **107**, 075420 (2023). https://doi.org/10.1103/PhysRevB.107.075420.

[8.21] S. Shrestha, C. S. Chang, S. Lee, N. L. Kothalawala, D. Y. Kim, M. Minola, J. Kim, and A. Seo, "Raman spectroscopic characterizations of graphene on oxide substrates for remote epitaxy", *J. Appl. Phys.* **133**, 105301 (2023). https://doi.org/10.1063/5.0143083.

[8.22] V. O. Mercadillo, H. Ijije, L. Chaplin, I. A. Kinloch, and M. A. Bissett, "Novel techniques for characterising graphene nanoplatelets using Raman spectroscopy and machine learning", *2D Mater.* **10**, 025018 (2023). https://doi.org/10.1088/2053-1583/acc080.

CHAPTER 9

Ferroelectric Oxides and Others

9.1 DIFFERENCE RAMAN SPECTRA OF PbTiO₃ FILMS

We have performed Raman scattering investigations for ferroelectric $PbTiO_3$ thin films grown on both single-crystal $KTaO_3$ and fused quartz substrates by means of metal-organic chemical vapor deposition (MOCVD) [9.1, 9.2]. The "difference Raman" technique was employed in which substrate contributions were subtracted to obtain Raman spectra for the $PbTiO_3$ films. Other Raman scattering studies on $PbTiO_3$ can be found in [9.3–9.5]. Lead titanate ($PbTiO_3$) is of considerable interest from both fundamental and practical points of view since it is characterized by a large spontaneous polarization, a small coercive field, and a high Curie temperature of ~500°C, and therefore, is a material with potential applications in both electronic and optical devices. $PbTiO_3$ thin films can be grown on a variety of substrates, including Si, $KTaO_3$, and $SrTiO_3$ by MOCVD methods, and on MgO substrates by rf-magnetron sputtering [9.1]. Ferroelectric thin films have been used for various applications such as nonvolatile memories, actuators, and sensors. It is known that electrical properties and Curie temperature of ferroelectric materials are significantly affected by lattice strain. For example, the Curie temperature of $PbTiO_3$ (PT) is decreased when hydrostatic pressure is applied and increased under two-dimensional strain [9.5].

Figure 9.1 illustrates the Raman spectra at 300 and 80 K for (a) and (a1) $PbTiO_3/KTaO_3$, (b) and (b1) a bare region of the $KTaO_3$ substrate, and (c) and (c1) for the difference spectrum between (a) and (b) and (a1) and (b1), respectively.

Figure 9.1(a) and (a1) show that the spectra from the $PbTiO_3$ region, while Figure 9.1(b) and (b1) resembled the Raman spectra characteristic of pure bulk $KTaO_3$. Spectra in Figure 9.1(a) and (b) are quite similar, and spectra in Figure 9.1(a1) and (b1) are also very similar, in band shapes and frequencies. However, the difference spectra in Figure 9.1(c) and (c1), clearly show different features, that is, the spectral features purely for $PbTiO_3$.

In Figure 9.1(c), the first four major bands located at 86, 146, 207, and 284 cm⁻¹, marked 1, 2, 3, and 4, respectively, are Raman characteristic modes from pure $PbTiO_3$ at 300 K. Two relatively strong bands at 466 and 591 cm⁻¹, are still due to $KTaO_3$, and their presence

DOI: 10.1201/9781032644912-9

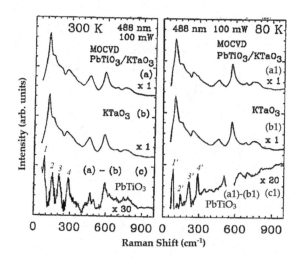

FIGURE 9.1   Raman spectra at 300 K (left) and 80 K (right) under excitation of 488 nm and 100 mW, (a) and (a1) from a PbTiO$_3$ film on KTaO$_3$, (b) and (b1) from the bare KTaO$_3$, (c) and (c1) the difference spectrum obtained by subtracting (b) from (a) [subtracting (b1) from (a1)] and expanded by a factor of 30 (and 20) with respect to (a), (b), (a1), and (b1), respectively.

Source: **From [9.1], figures 1 and 2 with reproduction permission of AIP.**

may arise from an effect of the dispersion dependence of absorption in the PbTiO$_3$ film on wave number. Similarly, from Figure 9.1(c1), the first four major bands located at 86, 147, 216, and 292 cm^{-1}, marked 1', 2', 3', and 4', respectively, are Raman modes from pure PbTiO$_3$ at 80 K. The difference spectra show the well-known soft modes of 1 and 1' both at 86 cm^{-1}, in Figure 9.1(c) and (c1), both of which are not resolved in the combined PbTiO$_3$/ KTaO$_3$ spectra between Figure 1(a)-(b), and between Figure 1(a1) and (b1). However, they are clearly displayed from the difference spectra of Figure 9.1(c) and (c1), at the same frequency at both 300 and 80 K.

It is interesting to observe that with a decrease of temperature from 300 K to 80 K, peak 2 shifts from 146 cm^{-1} up to 147 cm^{-1} (peak 2'), peak 3 shifts from 207 cm^{-1} up to 216 cm^{-1} (peak 3'), and peak 4 shifts from 284 cm^{-1} up to 292 cm^{-1} (peak 4'). This temperature dependence, which is contrary to that observed for many semiconductors and other solids, is characteristic of soft-mode behavior. According to the soft-mode theory, the temperature (T) dependence of the mode frequency ω can be described by the relationship,

$$\omega = \omega_0 \left| T - T_0 \right|^{1/2} \tag{9.1}$$

where ω_0 is the mode frequency at the transition temperature T_0. The T_0 characteristic of bulk PbTiO$_3$ is near 773 K and, thus, for temperatures between 300 and 80 K, Eq. (9.1) could be as:

$$\omega = \omega_0 \left(T_0 - T \right)^{1/2} \tag{9.2}$$

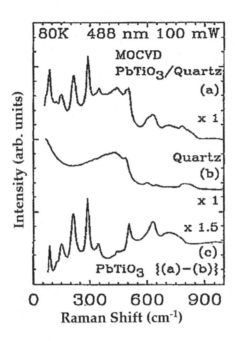

FIGURE 9.2 Raman spectra taken at 300 K under excitation of 488 nm and 100 mW, (a) from a PbTiO$_3$ film on quartz, (b) from the bare quartz region, (c) the difference spectrum obtained by subtracting (b) from (a) and expanded by a factor of 1.5 with respect to (a) and (b).

Source: From [9.1], figure 3, with reproduction permission of AIP.

As the temperature is decreased from 300 to 80 K, the mode frequency is expected to shift toward higher wave numbers, which is the case in Figure 9.1. Thus, the T-dependence of Raman spectra from PbTiO$_3$ at 300 K and 80 K are accounted for by the soft-mode theory [9.1].

PbTiO$_3$ film grown on quartz by MOCVD was investigated. Figure 9.2 presents Raman spectra, taken at 80 K and under 488 nm excitation of (a) PbTiO$_3$ thin film on a quartz substrate, (b) the quartz alone, and (c) the difference Raman spectrum [9.1]. Since the Raman spectrum characteristic of quartz is much broader than that of the PbTiO$_3$ film, the PbTiO$_3$ features can be clearly seen even in the case of Figure 9.2(a). The difference spectrum in Figure 9.2(c) shows that Raman bands at 88, 130, 151, 213, 288, 347, 446, 506, and 613 cm^{-1}, a broadband ranging from 670 to 800 cm^{-1}, and a peak near 720 cm^{-1}, all of which are features characteristic of crystalline PbTiO$_3$. The Raman spectra obtained at 300 K for this specimen (not shown here) exhibited features that are like those shown in Figure 9.2 for T = 80 K. As in the case of PbTiO$_3$ films on KTaO$_3$, the spectral line positions at 300 K for PbTiO$_3$ are somewhat lower than the values at 80 K [9.1].

9.2 RAMAN SCATTERING OF (Pb$_{1-x}$La$_x$)TiO$_3$ THIN FILMS

We have conducted Raman scattering studies of lead lanthanum titanate (Pb$_{1-x}$La$_x$)TiO$_3$ (PLT) thin film materials [9.6–9.8]. Similar research is found in [9.9, 9.10]. Lead lanthanum titanate (Pb$_{1-x}$La$_x$)TiO$_3$ (PLT) is an important ferroelectric ceramic material, and also a necessary partner of lead lanthanum zirconate titanate and lead lanthanum stannate zircinate titanate (PLZST). These PLT and related materials possess various novel properties, such as piezoelectricity, pyroelectricity, elasto-optic effect, linear, or quadratic electro-optic effect, which are useful in applications for nonvolatile memory devices, detectors, sensors, and optical switches [9.8].

Figure 9.3 exhibits Raman spectra taken at 80 K and under 457.9 nm excitation of $(Pb_{1-x}La_x)TiO_3$ for x = 0–0.18. The most intense lines observed at 522 cm^{-1} are from the Si substrate. The broad bands between 900 and 1150 cm^{-1} are due to second-order scattering from the Si phonon modes. As seen in Figure 9.3, all the TO-like modes (i.e., E(TO) or A_1(TO) decrease with increasing x(La), while all the longitudinal optical (LO)-like modes increase. Figure 9.4 exhibits Raman shift variations of various $(Pb_{1-x}La_x)TiO_3$ (PLT) modes with x(La) varying from 0 to 0.18, at 300 and 80 K.

As illustrated, all the modes except for the so-called "silent mode" show some systematic change as a function of x and T. The silent mode, marked in Figures 9.3 and 9.4, is neither Raman nor infrared active and exhibits a unique compositional and temperature dependence. Its frequency (287–288 cm^{-1}) does not vary with either varying x or T. The appearance of this feature may be attributed to the presence of disorder in the samples or an interface phase.

We have performed Raman scattering and related investigation for a series of highly textured lead lanthanum titanate $(Pb_{1-x}La_x)TiO_3$ (PLT) thin films grown on fused quartz substrates by MOCVD, with different x(La) of 0, 0.054, 0.167, 0.212, and

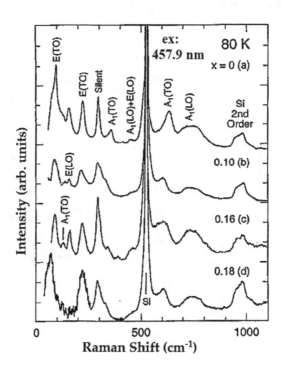

FIGURE 9.3   Raman spectra taken at 80 K and under 457.9 nm excitation of $(Pb_{1-x}La_x)TiO_3$ for x values of (a) 0, (b) 0.1, (c) 0.16, and (d) 0.18, respectively.

Source: **From [9.6], figure 3 with reproduction permission of AIP.**

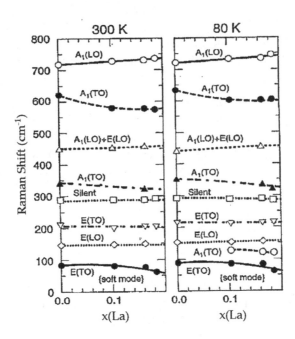

FIGURE 9.4   Raman shift variations of various $(Pb_{1-x}La_x)TiO_3$ (PLT) modes with x(La) varying from 0 to 0.18, at 300 and 80 K.

Source: From [9.6], figure 4 with reproduction permission of AIP.

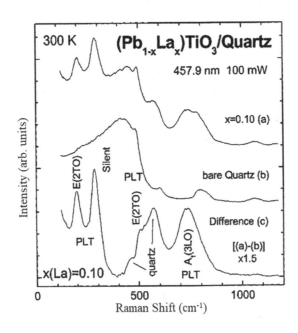

FIGURE 9.5   Raman spectra (300 K) under excitation of 4579Å and 100mW, from (a) x = 0.10 $(Pb_{1-x}La_x)TiO_3$ on quartz, (b) bare quartz region, and (c) difference of (a–b).

Source: From [9.8], figure 3 with reproduction permission of Elsevier.

0.32 [9.8]. Figure 9.5 shows Raman spectra at 300 K and under excitation of 4579 Å and 100 mW, from (a) x = 0.10 $(Pb_{1-x}La_x)TiO_3$ on quartz, (b) bare quartz region, and (c) difference of (a–b).

Raman spectra, measured at 300 and 80 K, showed the features of the PLT film and quartz substrate. By using a "difference Raman" technique, more PLT modes are shown. The variations of the PLT Raman modes with the La composition and the measurement temperature are studied, and related physical phenomena and problems are discussed. As shown in Figure 9.5, the E(2TO) mode at 197 cm^{-1} and the silent mode at 283 cm^{-1} are seen directly from the Raman spectrum without subtracting the contributions from the substrate. They are superposed upon the low-frequency side of a broad Raman band, peaking near 400 cm^{-1} from quartz. Beyond 400 cm^{-1}, there exists a weak peak at ~600 cm^{-1} and a band at 800 cm^{-1} from the quartz substrate in Figure 9.5(b). In the difference spectrum of Figure 9.5(c), two PLT features are observed: E(2TO) as a small bump at ~500 cm^{-1} and $A_1(3LO)$ band beyond 700 cm^{-1} [9.8].

9.3 RAMAN SCATTERING OF LEAD ZIRCONATE TITANATE (PZT)

Raman scattering studies of lead zirconate titanate (PZT) were performed by our collaborative team [9.11] and others [9.12–9.15]. $Pb(Zr_xTi_{1-x})O_3$ (PZT) is a useful material with excellent piezoelectric and ferroelectric properties, with many applications, including pyroelectric infrared sensors, electro-optical sensors, surface acoustic wave filters, microelectro-mechanical systems, and nonvolatile ferroelectric random access memory devices. PZT has a perovskite structure, displaying a tetragonal symmetry with the P4mm space group of C_{4v}. Tetragonal PZT in the P4mm space group yields eight Raman-active modes $(3A_1 + B_1 + 4E)$ [9.13].

In [9.11], thin films of lead zirconate-lead titanate $(PbZrO_3\text{-}PbTiO_3)$ of stoichiometry $Pb_1Zr_{02}Ti_{08}O_3$ were deposited on Pt and Si_3N_4/Si by the sol–gel process and characterized by techniques of Auger electron spectroscopy (AES), X-ray photoelectron spectroscopy (XPS), Raman spectroscopy, and Rutherford backscattering spectroscopy.

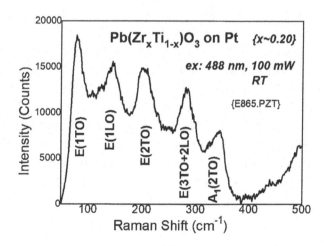

FIGURE 9.6   An original/unpublished Raman spectrum of a (PZT) with x(Zr) ~ 0.20.

Figure 9.6 presents an original/unpublished Raman spectrum measured at room temperature (RT) (300 K) and under 488 nm excitation, of a $Pb(Zr_xTi_{1-x})O_3$ (PZT) with $x(Zr) \sim 0.20$. Several Raman peaks are assigned: E(1TO), E(1LO), E(2TO), E(3TO+2LO), and A_1(2TO), at 77, 145, 200, 281, and 343 cm^{-1}, respectively. These peak assignments are following [9.11–9.14]. The lowest frequency E(1TO) mode, peaked at 77 cm^{-1}, is the so-called soft mode, like those in $PbTiO_3$ (PT) and $(Pb_{1-x}La_x)TiO_3$ (PLT) discussed in Sections 9.1 and 9.2. The temperature behavior of this soft-mode frequency shifts upward with decreasing temperature in accordance with the soft-mode theory and Eq. (9.2) like PT and PLT [9.11].

9.4 RAMAN SCATTERING OF BiFeO$_3$ FILMS

The perovskite bismuth ferrite $BiFeO_3$ (BFO) thin films were studied by Raman scattering and related techniques from our collaborative team [9.16] and others [9.17–9.20]. BFO not only has high polarization ($P_r = 60$–140 μC/cm^2) but also has a relatively small band gap ($E_g = 2.31$–2.67 eV), so that BFO can absorb a relatively wide solar spectrum and can obtain relatively high photoelectric conversion efficiency [9.16, 9.19]. Various methods have been used to improve the ferroelectric and photovoltaic properties of BFO thin films. It is reported that many factors such as doping, annealing atmosphere, and substrate all affect the photovoltaic performance of BFO thin films. The low-cost sol–gel method for the preparation of BFO films at high annealing temperatures has been reported, but when the film is annealed at too high temperature, it will form more abundant defects and bound space charges, resulting in lower remnant polarization and larger leakage current [9.16 and references therein].

We have prepared BFO thin films on SnO_2:F (FTO) substrates by the sol–gel method and rapidly annealed them at various temperatures and investigated them via X-ray diffraction (XRD), scanning electron microscopy (SEM), Raman spectroscopy, and XPS. The relationship between the microstructure and electric, optical, and photovoltaic properties was studied. The XRD, SEM, and Raman results show that a pure-phase BFO film with good crystallinity is obtained at a low annealing temperature of 450°C. Figure 9.7 shows the Raman spectra, at RT and under 514 nm excitation of BFO thin films annealed at 450°C, 500°C, and 550°C, respectively [9.16].

BFO material belongs to rhombohedral (R3c) in the space group, consisting of 13 Raman-active vibration modes of $4A_1 + 9E$. The optical modes of A_1 and E are marked in Figure 9.7, E(TO1), E(TO2), A1(TO1), E(TO3), E(TO4), E(TO5), E(TO6), E(TO7), E(TO8), E(TO9), and A_1(TO4), for peaks at 74, 138, 171, 217, 261, 275, 346, 372, 429, 470, and 520 cm^{-1}, respectively [9.16–9.20]. They are attributed to pure BFO Raman peaks. As the annealing temperature increases, their intensities increase, and the number of Raman modes increases. The Raman, XRD, and SEM results revealed that the BFO film annealing at 550°C has better crystallinity and film quality in this set of BFO films.

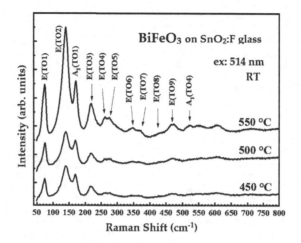

FIGURE 9.7   Raman spectra at RT and with 514 nm excitation of BiFeO$_3$ thin films on SnO$_2$:F (FTO) glass and annealed at various temperatures in an air atmosphere.

Source: From [9.16], figure 2, with reproduction permission from an open-access article of MDPI Materials.

9.5 HIGH-PRESSURE RAMAN SPECTROSCOPY OF Re$_3$N CRYSTALS

We have conducted a high-pressure Raman spectroscopy investigation on Re$_3$N crystals [9.21]. In [9.22, 9.23], the vibrational properties of Re$_3$N were studied from experiment and theory. The Re$_3$N samples in our study were synthesized through the high-pressure solid-state metathesis reaction between sodium perrhenate (NaReO4) and boron nitride. Raman scattering spectra were collected at RT in a confocal Raman system using a solid-state laser (532 nm) and a triple grating monochromator with an attached charge-coupled device. The Raman laser beam was focused within ~1 μm on the surface of the sample. The Raman system was calibrated using the band at 520 cm^{-1} of a silicon wafer with an uncertainty of 0.5 cm^{-1} [9.21].

Figure 9.8 shows (a) high-pressure (0–20 GPa) Raman spectra, measured under 532 nm and at RT, of hexagonal Re$_3$N samples and (b) Raman phonon frequencies of Re$_3$N as a function of pressure. In Figure 9.8(a), the zero-pressure spectrum was collected by using a ×100 objective lens, and other high-pressure spectra were collected by using a ×20 objective lens because of the limited height of the diamond anvil cell. In Figure 9.8(b), the open triangles, circles, and diamonds represent the decompression process to each mode of Re$_3$N, respectively. The solid lines are the least-square fit to the data using a two-order polynomial, and the dash lines are the calculated phonon frequencies.

Hexagonal Re$_3$N with four atoms in the unit cell, which generate 12 phonon branches containing 3 acoustic modes and 9 optical modes and has 6 vibrational modes in the long-wavelength limit: $\Gamma = 2E' + 2A_2' + E'' + A_1'$, and 4 optic modes are Raman

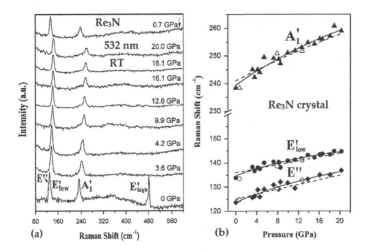

FIGURE 9.8 (a) High-pressure (0–20 GPa) Raman spectra, measured under 532 nm and at RT, of hexagonal Re$_3$N samples; the down arrow indicates decompression. (b) Raman phonon frequencies of Re$_3$N as a function of pressure.

Source: From [9.21], figure 3 with reproduction permission of Elsevier.

active: $\Gamma_{\text{Raman}} = 2E' + E'' + A_1'$. At zero-pressure of Figure 9.8(a), four Raman-active modes, E'', $E'_{(\text{low})}$, A_1', and $E'_{(\text{high})}$, are observed at 126, 129, 233, and 480 cm^{-1}, respectively. The low-frequency E'' and $E'_{(\text{low})}$ modes involve atomic vibrations of both Re and N atoms, mainly dominated by Re atom motion. The mode A_1' includes atomic vibrations from N atoms while Re atom does not vibrate along the directions x, y, and z because the N atom is much lighter than the Re atom, and has strong N–N covalent bond, which leads to relatively weaker electron–phonon interactions. The high-frequency mode $E'_{(\text{high})}$ at 480 cm^{-1} is mainly dominated by the N atoms motion. The well-crystallized Re$_3$N crystals exhibit strong orientation along c-axis of hexagonal symmetry leading to the disappearance of the mode $E'_{(\text{high})}$ at 480 cm^{-1}, as applied high pressure (3.6–20 GPa). Figure 9.8(a) shows the representative high-pressure Raman spectra of hexagonal Re$_3$N up to 20 GPa. Within creasing pressure, the three Raman modes of E'', $E'_{(\text{low})}$ and A_1' shift to higher frequencies and the line widths vary from 5.6 to 9.8 cm^{-1} without any sign of phase transition. The pressure-induced peak shifts of the vibrational modes confirm the mechanical stability of Re$_3$N.

Within experimental errors, the pressure dependence of the mode frequencies can be described by the second-order polynomial $\omega(P) = \omega_0 + aP + bP^2$, represented with solid lines in Figure 9.8(b). Where frequency ω is in cm^{-1} and pressure P in GPa. ω_0 is zero-pressure Raman phonon frequency, a and b being the first-order and second-order pressure coefficients, respectively. Pressure coefficient a is expressed as $(d\omega/dP)_{P=0}$, which is useful for determining zero-pressure mode–Grüneisen parameter

$$\gamma_0 = \left(\frac{d/n\omega}{d/nV} \right)_{P=0} = \frac{B_0}{\omega_0} \left(\frac{d\omega}{dP} \right)_{P=0} \tag{9.3}$$

where V is the molar volume and B_0 is the zero-pressure bulk modulus. The mode–Grüneisen parameters γ_0 are between 2.09 and 2.18 for the E'', $E'_{(low)}$, and A'_1 modes of Re_3N. γ_0 is considered a measure of anharmonicity, and it increases when the bonding force decreases. The larger mode–Grüneisen parameters in our Re_3N crystals reflect the instability of Re–N bonds under high pressure, which can be attributed to the spatial splitting of 5d orbitals of Re atoms or a strong hybridization between N 2p and Re 5d states, indicating a mixture of covalent and ionic bonding between Re and N atoms. The weakening of three-dimensional Re–N bonds can be correlated to iconicity or vacancy-related crystal defects resulting from the synthetic conditions under high pressure. Thus, we can get a tentative conclusion that our synthesized Re_3N crystals are between covalent and ionic crystals [9.21].

In Figure 9.8(b) the dash lines are the calculated Raman phonon frequency dependencies of pressure, based on density-functional theory and using Wu–Cohen exchange potentials [9.21, 9.22]. The calculated density of electronic states at the Fermi level shows that Re_3N is metallic in the pressure range of 0–20 GPa [9.22].

REFERENCES

[9.1] Z. C. Feng, B. S. Kwak, A. Erbil, and L. A. Boatner, "Difference Raman spectra of $PbTiO_3$ thin films grown by metalorganic chemical vapor deposition", Appl. Phys. Lett. **62**, 349–351 (1993). https://doi.org/10.1063/1.108954.

[9.2] Z. C. Feng, B. S. Kwak, A. Erbil, and L. A. Boatner, "Raman spectra of MOCVD-grown ferroelectric $PbTiO_3$ thin films", Mater. Res. Soc. Symp. Proc. **335**, 75–80 (1993).

[9.3] A. Bartasyte, S. Margueron, J. Santiso, J. Hlinka, E. Simon, I. Gregora, O. Chaix-Pluchery, J. Kreisel, C. Jimenez, F. Weiss, V. Kubilius, and A. Abrutis, "Domain structure and Raman modes in PbTiO3", Ph. Transit. **84**, 509–520 (2011). http://dx.doi.org/10.1080/01411594.2011.552433.

[9.4] X. L. Zhang, J. J. Zhu, G. S. Xu, J. Z. Zhang, L. P. Xu, Z. G. Hu, and J. H. Chu, "Temperature dependent spectroscopic ellipsometry and Raman scattering of $PbTiO_3$-based relaxor ferroelectric single crystals around MPB region", Opt. Mater. Expr. **5**, 2478 (2015). https://doi.org/10.1364/OME.5.002478.

[9.5] M. Nishide, T. Katoda, H. Funakubo, and K. Nishida, "Evaluation of strain components in PbTiO3 thin films by micro-Raman spectroscopy", J. Ceram. Soc. Jpn. **126**, 936–939 (2018). http://doi.org/10.2109/jcersj2.18149.

[9.6] Z. C. Feng, B. S. Kwak, A. Erbil, and L. A. Boatner, "Raman scattering and X-ray diffraction of highly-textured $(Pb_{1-x}La_x)TiO_3$ thin films", Appl. Phys. Lett. **64**, 2350–2352 (1994). https://doi.org/10.1063/1.111611.

[9.7] H. Y. Chen, J. Lin, K. L. Tan, and Z. C. Feng, "Characterization of lead lanthanum titanate thin films grown on fused quartz using MOCVD", Thin Solid Films **289**, 59–64 (1996). https://doi.org/10.1016/S0040-6090(96)08874-8.

[9.8] Z. C. Feng, J. H. Chen, J. Zhao, T. R. Yang, and A. Erbil, "Raman scattering of ferroelectric lead lanthanum titanate thin films grown on fused quartz by metalorganic chemical vapor deposition", Ceram. Int. **30**, 1561–1564 (2004). https://doi.org/10.1016/j.ceramint.2003.12.098.

[9.9] S. Bhaskar, S. B. Majumder, and R. S. Katiyar, "Diffuse phase transition and relaxor behavior in $(PbLa)TiO_3$ thin films", Appl. Phys. Lett. **80**, 3997 (2002). http://dx.doi.org/10.1063/1.1481981.

[9.10] G. F. G. Freitas, R. S. Nasar, M. Cerqueira, D. M. A. Melo, E. Longo, P. S. Pizani, and J. A. Varela, "Luminescence effect in amorphous PLT", J. Eur. Ceram. Soc. **25**, 1175–1181 (2005). http://doi.org/10.1016/j.jeurceramsoc.2004.04.008.

[9.11] M. J. Bozack, J. R. Williams, J. M. Ferraro, Z. C. Feng, and R. E. Jones, Jr., "Physical characterization of $Pb_1Zr_{0.2}Ti_{0.8}O_3$ prepared by the sol-gel process", J. Electrochem. Soc. **142**, 485–491 (1995). https://doi.org/10.1149/1.2044078.

[9.12] E. Buixaderas, I. Gregora, M. Savinov, J. Hlinka, Li Jin, D. Damjanovic, and B. Malic, "Compositional behavior of Raman-active phonons in $Pb(Zr_{1-x}Ti_x)O_3$ ceramics", Phys. Rev. B **91**, 014104 (2015). https://doi.org/10.1103/PhysRevB.91.014104.

[9.13] H. Fukushima, D. Ichinose, H. Funakubo, H. Uchida, H. Shima, and K. Nishida, "Angular dependence of Raman spectrum for $Pb(Zr,Ti)O_3$ epitaxial films", Jpn. J. Appl. Phys. **55**, 10TC07 (2016). http://doi.org/10.7567/JJAP.55.10TC07.

[9.14] B. Beklešovas, A. Iljinas, V. Stankus, J. Cyviene, M. Andrulevicius, M. Ivanov, and J. Banys, "Structural, morphologic, and ferroelectric properties of PZT films deposited through layer-by-layer reactive DC magnetron sputtering", Coatings **12**, 717 (2022). https://doi.org/10.3390/coatings12060717.

[9.15] M. Płonska and J. Plewa, "Investigation of praseodymium ions dopant on 9/65/35 PLZT ceramics' behaviors, prepared by the gel-combustion route", Materials **16**, 7498 (2023). https://doi.org/10.3390/ma16237498.

[9.16] J. Wang, L. Luo, C. Han, R. Yun, X. Li, X. Tang, Y. Zhu, Z. Nie, W. Zhao, and Z. C. Feng, "The microstructure, ferroelectric, optical and photovoltaic properties of $BiFeO_3$ thin films prepared by low temperature sol-gel method", Materials **12**, 1444 (2019). https://doi.org/10.3390/ma12091444.

[9.17] M. S. Bernardo, D. G. Calatayud, T. Jardiel, D. Makovec, M. Peiteado, and A. C. Caballero, "Titanium doping of $BiFeO_3$ ceramics and identification of minor phases by Raman spectroscopy", J. Raman Spectrosc. **48**, 884–890 (2017). https://doi.org/10.1002/jrs.5116.

[9.18] Z. Adineh and A. Gholizadeh, "Hydrothermal synthesis of Ce/Zr co-substituted $BiFeO_3$: R3c-to-P4mm phase transition and enhanced room temperature ferromagnetism", J. Mater. Sci. Mater. Electron. **32**, 26929–26943 (2021). https://doi.org/10.1007/s10854-021-07067-y.

[9.19] M. Benyoussef, S. Saitzek, N. S. Rajput, M. Courty, M. E. Marssi, and M. Jouiad, "Experimental and theoretical investigations of low-dimensional $BiFeO_3$ system for photocatalytic applications", Catalysts **12**, 215 (2022). https://doi.org/10.3390/catal12020215.

[9.20] M. Eledath and M. Chandran, "Raman study of $BiFeO_3$ nanostructures using different excitation wavelengths: Effects of crystallite size on vibrational modes", J. Phys. Chem. Solids **172**, 111060 (2023). https://doi.org/10.1016/j.jpcs.2022.111060.

[9.21] X. Jiang, L. Lei, Q. Hu, Z. C. Feng, and D. He, "High-pressure Raman spectroscopy of Re_3N crystals", Solid State Commun. **201**, 107–110 (2015). https://doi.org/10.1016/j.ssc.2014.10.031.

[9.22] A. Friedrich, B. Winkler, K. Refson, and V. Milman, "Vibrational properties of Re_3N from experiment and theory", Phys. Rev. B **82**, 224106 (2010). https://doi.org/10.1103/PhysRevB.82.224106.

[9.23] E. Deligoz, K. Colakoglu, H. B. Ozisik, and Y. O. Ciftci, "Vibrational properties of Re_2N and Re_3N compounds", Solid State Commun. **151**, 1122–1127 (2011). https://doi.org/10.1016/j.ssc.2011.05.028.

Index

Note: Page numbers in *italics* indicate a figure on the corresponding page.

Printed in the United States
by Baker & Taylor Publisher Services